ZAFAR ALAM MAHMOOD

L-LYSINE

ZAFAR ALAM MAHMOOD

L-LYSINE

L-Lysine - Production Through Fermentation

VDM Verlag Dr. Müller

Impressum/Imprint (nur für Deutschland/ only for Germany)

Bibliografische Information der Deutschen Nationalbibliothek: Die Deutsche Nationalbibliothek verzeichnet diese Publikation in der Deutschen Nationalbibliografie; detaillierte bibliografische Daten sind im Internet über http://dnb.d-nb.de abrufbar.

Alle in diesem Buch genannten Marken und Produktnamen unterliegen warenzeichen-, marken- oder patentrechtlichem Schutz bzw. sind Warenzeichen oder eingetragene Warenzeichen der jeweiligen Inhaber. Die Wiedergabe von Marken, Produktnamen, Gebrauchsnamen, Handelsnamen, Warenbezeichnungen u.s.w. in diesem Werk berechtigt auch ohne besondere Kennzeichnung nicht zu der Annahme, dass solche Namen im Sinne der Warenzeichen- und Markenschutzgesetzgebung als frei zu betrachten wären und daher von jedermann benutzt werden dürften.

Coverbild: www.purestockx.com

Verlag: VDM Verlag Dr. Müller Aktiengesellschaft & Co. KG
Dudweiler Landstr. 99, 66123 Saarbrücken, Deutschland
Telefon +49 681 9100-698, Telefax +49 681 9100-988, Email: info@vdm-verlag.de

Herstellung in Deutschland:
Schaltungsdienst Lange o.H.G., Berlin
Books on Demand GmbH, Norderstedt
Reha GmbH, Saarbrücken
Amazon Distribution GmbH, Leipzig
ISBN: 978-3-639-23151-9

Imprint (only for USA, GB)

Bibliographic information published by the Deutsche Nationalbibliothek: The Deutsche Nationalbibliothek lists this publication in the Deutsche Nationalbibliografie; detailed bibliographic data are available in the Internet at http://dnb.d-nb.de .

Any brand names and product names mentioned in this book are subject to trademark, brand or patent protection and are trademarks or registered trademarks of their respective holders. The use of brand names, product names, common names, trade names, product descriptions etc. even without a particular marking in this works is in no way to be construed to mean that such names may be regarded as unrestricted in respect of trademark and brand protection legislation and could thus be used by anyone.

Cover image: www.purestockx.com

Publisher:
VDM Verlag Dr. Müller Aktiengesellschaft & Co. KG
Dudweiler Landstr. 99, 66123 Saarbrücken, Germany
Phone +49 681 9100-698, Fax +49 681 9100-988, Email: info@vdm-publishing.com

Printed in the U.S.A.
Printed in the U.K. by (see last page)
ISBN: 978-3-639-23151-9

L-LYSINE

PRODUCTION THROUGH FERMENTATION

BY

DR. ZAFAR ALAM MAHMOOD

PREFACE

L-lysine is one of the well known essential amino acid with great demand and application in pharmaceuticals and as an additive to animal feed or human food-stuffs. The importance of L-lysine as an essential amino acid in the nutrition of human beings has made it a desirable supplement of the diet in recent years. This is more prominent in underdeveloped and over-populated areas of the world, where the chief staples have been found deficient of this amino acid. The global demand of L-lysine has increased tremendously during the last three decades and is further expected to increase and for these reasons, efforts are now being focused on the potential of L-lysine derived bacteria and its application. This book is also compiled with the aim to provide comprehensive information on L-lysine fermentation along with the history and techniques reported year after year.

The production of L-lysine is achieved through fermentation of an auxotrophic – regulatory mutant developed from a locally isolated bacterial strain of *Corynebacterium glutamicum* and reported in detail along with the newer techniques, such as genetic manipulation / cloning for development of efficient L-lysine producing strains. Hope, the book will not only provide ample information for researchers working on isolation and identification of starins, especially relating to *Corynebacterium glutamicum* and on various techniques for improvement and development of an afficient L-lysine producing strain but will also lift the curtain to fill both a gap and a need in industrial production of L-lysine. The book also encompases, the modified classical techniques, such as "Bio-autographic Technique" and "Penicillin Selection Technique" followed by UV irradiation for the development of auxotrophic mutants as well as to produce and develop high yield L-lysine strains through mutant resistant to AEC and the Auxanographic technique for the evaluation of growth factor(s) requirements of the auxotrophic mutants. However, at the same time, apart from all these classical techniques, the recent advancement and techniques (genetic engineering / cloning) reported for the development of high yield L-lysine have also been reviewed and discussed.

I am pleased to dedicate this book to my parents in recognition of their efforts in developing and building my life, personality and skills. My gratitude goes to Professor Dr. S.M.S. Zoha, Professor Dr. Dilnawaz Shaikh, Department of Pharmaceutics, Faculty of Pharmacy, University of Karachi for their help and assistance. My thanks also to Professor Dr. M. Rafi Shaikh and Professor Dr. Usman Ghani Khan for their valuable advice. Thanks to my company, Colorcon Limited – UK, for support and encouragement in academic work. Finally, I would like to thank Mr. Benoit Novel, Director and Miss. Tabassum Dawlut, Acquisition Editor of VDM Publishing House, Germany for their inspiration and support.

DR. ZAFAR ALAM MAHMOOD

II

CONTENTS

LIST OF TABLES

TABLE No.   **PAGE**

LIST OF FIGURES

INTRODUCTION

Microorganisms are among man's best friends as well as worst enemies but it took him a million years to find it out (Stakman, 1964).Although the industrial application of microorganisms is believed to have originated in ancient times, with the production of fermented beverages and food stuffs, but until the time of Pasteur, there was no understanding of the fermentation process. It was Pasteur,who in 1857 discovered that fermentation is a characteristic of living organisms, and that spontaneous generation does not occur (Miller and Litsky, 1976). These studies were extended to many bacteria and filamentous fungi. It was observed that during the process a number of potentially useful products were excreted into the medium in which the organisms were grown. However, the scientific explanation of fermentation started only with the understanding of the role of microorganisms in such processes. In the early stages of fermentation studies, it was observed that fermentation is either growth related or related to a well defined substrate which was converted into a single product or multiple products. However, in both the cases the substrate served as a source of energy for microorganisms.

During the growth process of microorganisms, some secondary metabolites which were neither dependent on substrate concentration nor related to growth process, such as antibiotics, steroids, etc., led to a completely new concept of biotransformation of organic molecules. Continued research on the subject thus resulted in host of useful products, whereas, engineering advancements have allowed their large scale, economical production. This was the beginning of modern biotechnology. The search for new biological materials, to be used as drugs and pharmaceuticals, resulted in a phenomenal growth of industrial microbiology on one side and fermentation engineering on the other. The increased industrial utilization of biological processes suggest that biotechnology will be a major growth industry in near future and this will affect the lives and welfare of people all over the world. Apart from the role of microorganisms in the production of drugs and pharmaceuticals, it was observed that biotransformation could also be utilized for the production of food and feed materials. In the field of food, the role of microorganisms in the preservation of raw and cooked materials and improvement of flavours and colours was fully established. The food industry thus benefited most from the recent advances in the field of fermentation. However, with the establishment of the role of growth promoting substances, such as vitamins, amino acids and gibberellins, much attention was focussed on the utilization of microorganisms and their enzymes, to produce these substances. A large number of microbial strains were developed for the industrial production of these useful organics and here

1

again the help of fermentation process came to the advantage of mankind.

Over the last forty years, the knowledge of microorganisms has increased considerably and with it has grown the appreciation of its potentials. We are, therefore, at a point in the history of biotechnology where it is possible to indicate, with a fair degree of accuracy, whether or not a given substance could be produced microbiologically. Moreover, which type of organism has to be used; from where that particular strain might be isolated; under what conditions it has to be grown to express the desirable properties; and how its productivity might be improved, either by judicious choice of culturing conditions, or by addition or subtraction of a particular nutrient or by genetic engineering. This rational approach to biotechnology is now open for further exploitation. What is most striking about biotechnology, however, is that it has something to offer to both the industrially developed as well as the underdeveloped countries alike, irrespective, to some extent, of the abundance of natural resources.

The microbial production of amino acids has generated special interest throughout the world specially in recent years. The impetus for these advances originated chiefly from the interest in the nutritional application of amino acids, which have long been recognized in the nutrition of human beings and domestic animals, especially for single-stomach (monogastric) animals like broilers, poultry and swine (Osborne and Mendel, 1914 and 1919; Scrimshow and Altschul, 1971, Hirose and Shibai, 1985, Eggeling and Sahm, 1999, Tryfona and Bustard, 2005 and Anastassiadis, 2007). Nevertheless, the multifarious uses of amino acids, such as flavouring agents (Shallenberger et al., 1969), pharmaceutical (Kaneko and co-workers, 1974 and Anastassiadis, 2007) and as a raw materials in the chemical and cosmetic industries (Yamamoto, 1978 and Anastassiadis, 2007) have also played a significant role in the enhancement of the research activities in this particular field.

The systematic research on the synthesis of amino acids by microorganisms started during the late 1940's and it began to bear fruits towards the later part of the 1950's during which microbial synthesis of few amino acids became possible. The best example is of L-lysine (Mitchell and Houlahan, 1948; Windsor, 1951). L-lysine (2, 6-diaminohexanoic acid; alpha, epsilon-diaminocaproic acid – $C_6H_{14}N_2O_5$; MW 146.19 daltons, basic in nature, positively charged at physiological pH, highly water soluble, and pKa's: $pKa_1 = 2.20$, $pKa_2 = 8.90$ and $pKa_3 = 10.28$) is one of the well known essential amino acid, marketed as L-lysine monohydrochloride or as free base, L-lysine and has great demand in pharmaceutical, food, feed and cosmetic industries (Kaneko

2

and co-workers, 1974; Kagan, 1974; Chaitow, 1985; Tosaka and Takinami, 1986, Balch and Balch, 1990) and as an additive to animal feed or human food-stuff (Feldberg and Hetzel, 1958a, 1958b; DeMuelenaere et al., 1967; Tosaka and Takinami, 1986). The importance of L-lysine as an essential amino acid in the nutrition of human beings has made it a desirable supplement of the diet in recent years, particularly in the under-developed and over-populated areas of the world, where the chief staples have been found deficient in this amino acid. In addition, as reported earlier, L-lysine is also required as feed additive in the animal feed industry, commonly mixed with livestock (for example cereals which are deficient in L-lysine – an exception is wheat germ, which contains high amount of lysine) for poultry and pig breeding (Hirose and Shibai, 1985, Eggeling and Sahm, 1999, Tryfona and Bustard, 2005 and Anastassiadis, 2007).

L-lysine may be produced either by isolation from natural materials (originally from the hydrolysis of animal or plant proteins) or by chemical, microbial or enzymatic synthesis. Chemical synthesis produces a recemic (DL-lysine) product, which requires additional optical resolution, as the amino acid in the L-isomeric form is required in all its application. The microbial synthesis, however, gives rise to optically pure L-lysine, thus making the process advantageous over the synthetic one. The possibility of microbial production of L-lysine was first realized by the findings, that L-lysine can be synthesized from α-aminoadipic acid by *Yeast,* and *Neurospora* (Mitchell and Houlahan, 1948; Windsor, 1951) or from diaminopimelic acid (DAP) by *Escherichia coli* (Dewey and Work,1952; Davis, 1952). The bacterial production of L-lysine through the decarboxylation of DAP (Casida and Baldwin, 1956) by Chas Pfizer and Company Inc., in the United States provided the first example of the preparation of an amino acid by a microbial process on a commercial scale. However, very soon a modified process in which a single organism produced the DAP and converted it to L-lysine, was reported (Kita and Huang, 1958). The new process was an improvement of the previous method (Casida and Baldwin, 1956) but still retained the disadvantages of the two-step process, because the product was recovered only after the cells were lysed. With the original mutant reported by Casida and Baldwin (1956), it was necessary to use glycerol as a carbon source, to obtain high yields of DAP. Later a double auxotroph of *Escherichia coli*, which was deficient in L-histidine as well as in L-lysine, was isolated and found to give better yields of DAP from inexpensive source of carbohydrate, such as lactose, sucrose, and molasses (Huang et al., 1960).

The development of direct L-glutamic acid fermentation process, by an efficient glutamic acid producer, *Corynebacterium glutamicum* (Synonym *Micrococcus glutamicus*) was reported by Kinoshita et al. (1957a) at Kyowa Hakko Kogyo Company, Japan. This discovery had great economic impact in the field of L-lysine fermentation, as the successful commercialization of this work led to the production of L-lysine on large scale, using mutants of *Corynebacterium glutamicum* (Kinoshita et al., 1958b). The modified process has proven to be the most successful for the production of L-lysine, as well as, other amino acids (Nakayama et al., 1961a), since it has obvious advantages of a one-step process. The success of L-lysine production brought about a new concept in fermentative production of amino acids. The new production method used artificially derived auxotrophic or regulatory mutants which were resistant to growth inhibition by the structural analogues of the desired amino acid. The complex regulatory system, controlling the biosynthesis could be artificially by-passed, in these mutants to allow over production of desired product. Various amino acid can now be produced, using the corresponding auxotrophic and regulatory mutant strains. Based on these, the microorganisms used in the fermentative production of L-lysine or any other amino acid may be classify into four different groups, such as utilization of wild strain, auxotrophic mutants, regulatory mutants and a combination of both auxotrophic and regulatory mutants. Now a days, improved production of L-lysine by fermentation using *Corynebacterium glutamicum, Brevibacterium flavum, Brevibacterium lactofermentum* and some other organisms has been attained with both auxotrophic – regulatory mutants and mutants developed through genetic menipulation. (Bustard and Tryfona, 2005, Obeta and Ekwealor, 2006, Ikeda et al., 2006, Hayashi et al., 2006, Anastassiadis, 2007, Wittmann and Becker, 2007, Tateno et al., 2007, Blombach et al., 2009. and Bulfer et al., 2010).

In the fermentative production of L-lysine, all raw materials are natural or biologically available substances. No harmful by-products have been found in L-lysine fermentation. On the contrary, useful substances remain in the spent-broth from which many bye-products could be recovered. The spent-broth still contains various useful substances, including organic and inorganic nitrogen compounds, phosphorus compounds, and potassium salts which could be used as animal feed additive and fertilizer. Based on this, fermentative production of L-lysine as an animal feed additive was reported (Kircher and Pfefferle, 2001). The final product obtained through three different steps was reported as Biolys®60. The steps include, fermentation using a suitable strain of *Corynebacterium glutamicum* in starch hydrolysate or sucrose as the carbon source and ammonium sulphate and ammonia as the nitrogen source, followed by concentration of the main broth once the fermentation process is completed and then finally drying of the concentrated broth to produce Biolys®60 granules.

At present mass-production of amino acids is limited to L-lysine and L-glutamic acid and a few other. The world annual production data from 1975 to 2010 for L-lysine is given below. Production figure of 2009 indicates around 900,000 T with an expected growth of around 8 to 10% over the next couple of years (Nakayama, 1985, Wittmann, 2007, Connor, 2008 and Blombach, 2009).

Annual L-Lysine Production Throughout the World

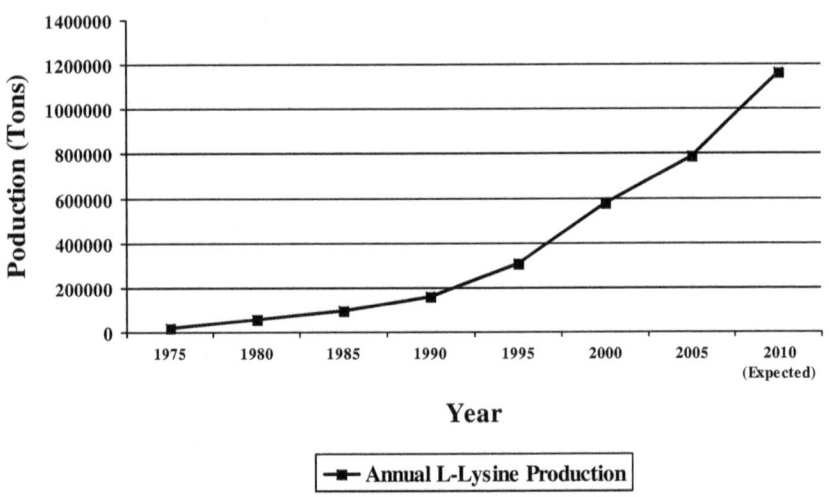

The Major producers of L-lysine in the world are: Ajinomoto Company (Japan) & AjinomotoEurolysine S.A.S (France), ADM (USA), Koyawa Hakko Kogyo Company (Japan), Sewon / Miwon (Korea), BASF (Germany), Degussa / Fermas (Slovakia). CJ Corporation (South Korea), Cargill Corporation (USA) and Fermex (Mexico) In terms of tonnage, L-lysine is one of the major commercial product made through fermentation. The geographic location of L-lysine production capacity as per 2002 statisistics revealed North America as the biggest producer of L-lysine (50%), followed by Asia - Japan & Korea (15%) and other Asian countries (20%), and Europe (15%) (Connor, 2008). Cheil's plant in Indonesia is also producing L-lysine. In addition, the worlds leading L-lysine producers themselves are increasingly investing in L-lysine plants in China,

Vietnam, Thailand and other South East Asian countries. The Chinese L-lysine production capacity has tremendusly increased in recent years. The 2007 figure indicates around 335,650 T.

The demand of L-lysine is still increasing in the field of foods, animal feeds, pharmaceuticals and chemicals, etc. To meet increasing and diversified demands, there is still room for improvement in strains based on the knowledge of microbial physiology. Furthermore, studies on process optimization, especially for lowering the expense of carbon and energy source is also desirable. In Pakistan, due to the well established animal feed, food, and pharmaceutical industries the demand for L-lysine has been very high. All of the L-lysine being used in animal feed, food and pharmaceutical industries is imported from overseas. It has been estimated that volume of L-lysine imported is about 800 tons per year. (n.d., 2009). Thus it will be highly beneficial in terms of balancing the country's payment as well as creating an opportunity for fermentation industry in Pakistan and other underdeveloped countries, if L-lysine production can be commercialized locally.

REVIEW OF THE LITERATURE

1. ORIGIN OF AMINO ACIDS

Amino acids are the chemical units or the "building blocks," that make up proteins. Proteins could not exist with out the proper combination of amino acids. In order for a protein to be complete, it must contain all of its particular amino acids. Amino acids contain about 16% nitrogen that distinguishes them from carbohydrates and fats in the body (Balch and Balch, 1990). The first amino acids were probably produced on the earth more than years ago via "prebiotic synthesis" in the primordial atmosphere (Kleemann et al., 1985). The concept of prebiotic synthesis reported to be based on laboratory experiments in which glycine, alanine, aspartic acid, glutamic acid, and other compounds were produced by the action of an electrical discharge on a simulated primordial atmosphere consisting of methane, hydrogen, water and ammonia. Since then traces of amino acids have been detected in moon rocks, meteorites and interstellar space (Kleemann et al., 1985).

2. DISCOVERY OF L-LYSINE

The observation that proteins, when subjected to the hydrolytic action of boiling acid, were decomposed into relatively simple crystalline substances - "the amino acids", was made more than a century ago. In 1806, two French investigators, Vauquelin and Robiquet, isolated asparagine from asparagus juice (Kleemann et al., 1985). This was the beginning of the history of amino acid chemistry. At first only a few amino acids were distinguished but as time went on more of them were isolated. It was not until 1925 that Schryver and Buston isolated threonine from oat protein, the last discovered of the twenty protein forming amino acids (Considine, 1974; Annexure - I). Although three hundred amino acids were so far discovered in nature, but only twenty of these were reported to be present in proteins from all forms of life plants, animals, or microbes (Rodwell, 1993). Ten of them, including L-lysine were considered essential as they were required to be present in diet (Chaitow, 1985).

The structure of the amino acids formed upon the hydrolysis of proteins has been established both by degradation and by synthesis in the course of extensive work carried out during 1850-1950. The simplest amino acid, glycine was the first to be recognized as a product of animal protein hydrolysis (Braconnot, 1820). Various methods were reported for the hydrolysis of proteins. Braconnot (1820) heated the animal protein with sulphuric acid, while Bopp (1849) used hydrochloric acid to hydrolyse proteins (albumin, casein and fibrin). Later, Hlasiwetz and Habermann (1873) reported that animal proteins could be effectively hydrolysed by means of

7

hydrochloric acid in presence of tin. The decomposition took place smoothly and the products obtained were identified as leucine, tyrosine, glutamic acid, aspartic acid and ammonia, with only a small residue of unknown material. Similarly Schutzenberger (1879) decomposed proteins (albumin) by heating them at 150°C with barium hydroxide and likewise obtained aspartic acid, glutamic acid, tyrosine and leucine with a series of substances reported as homologues of leucine. The whole problem of the completeness with which a protein could be accounted for in terms of its products of hydrolysis was discussed by Drechsel (1889), who reported the isolation of a basic compound from the mother liquor of the casein hydrolysate after separating the insoluble L-glutamic acid. Later the word "Lysatin" was as signed for this new compound (Drechsel, 1890). The presence of this basic compound was also reported in a variety of proteins, such as conglutin, glutenfibrin (gliadin) hemiprotein, oxyprotsulfonic acid and egg albumin (Siegfried, 1891).

Drechsel (1891), finally proposed the name "Lysine" to this basic compound with the suggestion that it might well be a diaminocaproic acid, the next higher homologue to ornithine, with which it closely resembled. In order to prove this, Drechsel and Kruger (1892) attempted to obtain pentamethylenediamine from lysine by decarboxylation, but failed. Hedin (1895), the discoverer of arginine reported that the basic compound, lysine (originally lysatin) was a mixture of lysine and arginine, the later component being the source of urea when the mixture was treated with hot barium hydroxide. The correctness of all views was demonstrated by subjecting lysine and ornithine to prolonged anaerobic putrefaction and benzoylation of the products (Ellinger,1899 and 1900). The products obtained by lysine and ornithine were dibenzoylcadaverine and dibenzoylputrescine (m.p. 178°C) respectively. Based on these findings, Ellinger (1899 and 1900) suggested that the most probable constitution of lysine could be expressed by a formula given below.

$NH_2CH_2-CH_2-CH_2-CH_2CH(NH_2)-COOH$

However, the position of the carboxyl group could not be established. Henderson (1900) pointed out that the carboxyl group could not be placed on the middle carbon atom because of the optical activity. The formation of acetic acid and propionic acid by the fusion of lysine with an alkali strongly supported the evidence that the carboxyl group was at the end of a straight chain. The synthesis of lysine by Fischer and Weigert (1902) from γ-cyanopropylmalonic ester by reacting it with nitrous acid followed by reduction, consequently elucidated the structure of lysine as α, ε-diaminocaproic acid.

8

$$CH_2\text{-}CH_2\text{-}CH_2\text{-}CH_2\text{-}CH_2\text{-}COOH$$
$$NH_2 \qquad\qquad NH_2$$

The crystalline form of lysine was finally obtained by Vickery and Leavenworth (1928).

3. PRODUCTION OF L-LYSINE BY MICROORGANISMS

3.1. L-lysine from naturally occurring microorganisms.

Microbial metabolites of nitrogenous origin was first recognized in the last century by Thenard, Pasteur, Duclaux and others in their studies on yeast fermentation (Kinoshita, 1959). The composition of many microbial cell proteins was analyzed to determine their content of essential amino acids. The possibility of the microbial production of lysine was rapidly realized by the findings that lysine is synthesized from α-aminoadipic acid by yeast and Neurospora (Mitchell and Houlahan, 1948; Windsor, 1951). Reusser et al. (1957) analysed 19 species, including bacteria, streptomyces and yeasts. In general, L-lysine content was recorded high, tryptophan and threonine content low and methionine content extremely low. The L-lysine content of many bacteria, yeast and molds have been studied and reported by various workers (Anderson and Jackson, 1958; Nelson et al., 1960; and Rhodes et al., 1961). In addition, algae as a source of Lysine have also been reported (Hundley and Ing, 1956). Richards and Haskins (1957) reported the extracellular lysine formation in 600 strains of fungi. They found that *Ustilago maydis* (PRL 1092 and 1229) and strains of *Gliocladium sp*. Had active lysine productivity. Detailed studied on PRL 1092 and PRL 1229 were done by Dulaney et al. (1956) and Dulaney (1957) who reported yields of free lysine, i.e., 200 to 300 mg/ml. Ericson and Kurz (1962) and Kurz and Ericson (1962) also screened a few *Ustilago sp*, including *Ustilago maydis* PRL 1092. However, they found that *Ustilago maydis* accumulated only 120 to 200 mg/ml of free lysine. Tauro et al. (1963) reported the screening results of few *Ustilaginales* fungi for free lysine accumulation. Among the five cultures of *Ustilaginales* screened, Sphacelotheca sorghi, produced the maximum amount, i.e., 50 mg/ml of free L-lysine in its culture medium. They also reported an increase in the free lysine content of culture broths of *Ustilago maydis* PRL 1092 upto 400 mg/ml. Alfredo et al., (1969) reported extracellular production of lysine by mutants of *Ustilago maydis* in agave (*Agave desertii*) juice. Upto 2.5 g L-lysine /L of broth was obtained with a UV-ethyleneimine combination mutant of *Ustilago maydis* in a culture medium containing agave juice as the main constituent. On the sugar consumption basis, 57% conversion was obtained. A very comprehensive review describing production of L-lysine by yeasts and the enzyme involved in its biosynthetic pathway was presented by Soda et al. (1981).

3.2 Diaminopimelic acid fermentation and its decarboxylation to produce L-lysine.

The bacterial production of L-lysine has been achieved through several pathways. Davis, 1952 and Dewey and Work, 1952 suggested to produce L-lysine via conversion of its immediate precursor, diaminopimelic acid (DAP). The DAP was first isolated from bacteria (Work, 1950 and 1951) and found to be decarboxylated to L-lysine and carbon dioxide by an extract of *Escherichia coli* (Davis, 1952; Dewey and Work, 1952). Later Casida and Baldwin (1956) was awarded a patent on a process in which a lysine auxotroph of *Escherichia coli* ATCC 12408 produced and accumulated DAP (9g/l) in a medium containing glycerol as the carbon source. A process utilizing only one organism to produce L-lysine was reported by Kita and Huang in 1958. This was the modification of the approach used by Casida and Baldwin (1956). A lysine requiring mutant of Escherichia coli was used as a single organism which produced the DAP and converted it to L-lysine, only after the cells have been lysed by organic solvent (e.g., benzene) or by sonic oscillation. This process was an improvement, but it retains the disadvantages of a two step process. The production of DAP was reported in a number of lysine auxotrophic cells, e.g., Escherichia coli (Angulo et al., 1960a and 1960b), Corynebacterium glutamicum (Nakayama and Kinoshita, 1961a and 1961b), Bacillus subtilus (Tanaka et al., 1967), Arthrobacter ureafaciens (Kase and Nakayama, 1970), Bravibacterium ammoniagenes (Hagino and Nakayama, 1970), Aeromonas formicans (Rhuland et al., 1955), Generally a mixture of meso and L-DAP formation has been reported, however, in some strains of Arthrobacter ureafaciens (Kase and Nakayama, 1970), only the L-lysine has been dectected. In Escherichia coli and Corynebacterium glutamicum, DAP accumulation was observed due to the deficiency in DAP-decarboxylase. The mutant which accumulated only the L-DAP may be deficient in DAP racemase (Nakayama, 1972a). Detection of DAP by paper chromatography (Rhuland et al., 1955) and by analytical methods (Work, 1957 and Gilvarg, 1958) has also been reported. Furthermore, a number of antimicrobial compounds have been prepared from DAP (Nakayama, 1972a), but as such there is no major demand for DAP since it cannot serve as a nutritional substitute for L-lysine. Although L-lysine can be produced from DAP as described earlier, the methods involve two steps and thus inferior and uneconomical.

3.2. Direct fermentation of L-lysine using auxotrophic and regulatory mutants.

The most important approach to L-lysine production received its impetus from work of Kinoshita et al. (1957a; 1957b; 1958a and 1958b). They developed a number of auxotrophic mutants of Corynebacterium glutamicum following ultraviolet (UV) or cobalt (Co^{60}) irradiation. The prototroph was a biotin requiring isolate found to possess an amazing capacity to produce glutamic acid in high quantities. The biosynthetic blocks, which were found to be most conducive for effective L-lysine accumulation were homoserine or methionine and threonine (Nakayama et al., 1961a). Later, Kinoshita et al., (1961) were granted a patent for production of L-lysine by auxotrophs of Micrococcus glutamicus (Later named Corynebacterium glutamicum) of the type described. This direct approach was proved to be quite successful for the production of L-lysine as well as other amino acids (Nakayama et al., 1961b), since it has the obvious advantages of a one step process. Similar process was reported with a homoserine auxotroph of Brevibacterium flavum (Sano and Shiio, 1967 and Shiio and Sano, 1969). Double auxotrophs, which require in addition to homoserine, at least one of the amino acids, threonine, isoleucine, or methionine, for growth, have been found to be highly stabilized, showing little tendency to revert to homoserine independence. It is possible not only to prevent reversion of the cultures to a wild type state, but also to produce L-lysine in higher yields since many of the microorganism are double mutants in the homoserine pathway (Kinoshita and Nakayama, 1978 and Nakayama, 1983). L-lysine production has been studied in homoserine auxotrophs by various investigators. Some important organisms are: *Brevibacterium ammoniagenes, Bacillus subtilus* (Kinoshita et al., 1961; Nakayama et al., 1961a and Nakayama and Kinoshita, 1961c), *Escherichia coli* (Nakayama et al., 1961a), Brevibacterium sp. (nakayama, 1972a; Kinoshitan et al., 1961; Sano and Shiio, 1967; Areshkina et al., 1965a and 1965b and Grivina, 1967), *Azotobacter suis* (Petrov et al., 1964; and 1965), *Brevibacterium flavum* (Sano and Shiio, 1967), *Corynebacterium acetophilum* (Seto and Harada, 1969), *Pseudomonas aeruginosa* (Kinoshita et al., 1961), Aerobacter aerogenes (Kinoshita et al., 1961), *Corynebacterium hydrocarboclastus, Brevibacterium ketoglutaricum* and *Arthrobacter paraffineus* (Nakayama, 1972a), *Corynebacterium glutamicum,* AECR-13 and AECR-25 (Misra et al., 1979 and 1980), *Corynebacterium glutamicum* ATCC 13286 and ATCC 21526 (Hallaert et al., 1987), *Corynebacterium glutamicum* 9366-AEC/100 Leu-6 (Fan et al., 1988), Bacillus sp. (Schendel et al., 1990), *Corynebacterium glutamicum* (Sur et al., 1991). Shiio and Sano (1969) 1solated a homoserine auxotroph of *Brevibacterium flavum* (S-20) having homoserine dehydrogenase activity. The strain produced 23g ofr L-lysine from 100g glucose. The growth was completely inhibited with L-methionine or L-threonine (10 µg/ml) and restored by higher concentration of these amino acids. Decreased homoserine dehydrogenase activity was the primary result of the mutation. Another homoserine auxotrophic mutant (H-1013) produced 34.1g L-lysine/L of broth. An important advancement in the production of L-lysine was the introduction of an analogue resistant strain. Sano and Shiio (1970) used S-(2-aminoethyl)-L-cysteine (AEC) resistant strains of *Brevibacterium flavum* 2247 for the production of L-lysine. Strain FA-3-115 produced 31.8g/L of L-lysine in a glucose (10%) medium while strain FA-1-30 produced 30.8g/L of L-lysine in molasses (13% as glucose) medium.

Hallaert et al. (1987) described production of L-lysine by two homoserine auxotrophs of *Corynebacterium glutamicum* resistant to AEC. *Corynebacterium glutamicum* ATCC 13286 was found to be homoserine auxotrophs, while strain ATCC 21526 was reported to be homoserine plus leucine auxotroph. Under submerged condition, *Corynebacterium glutamicum* ATCC 13286 produced 18g L-lysine /L of medium. Fan et al. (1988) reported a mutant strain of Brevibacterium flavum FML 8611 resistant to AEC and rifampicin. The strain produced 78.9g L-lysine /L of medium in 80 hours. In a similar type of study, Plachy et al. (1988) also reported isolation of a mutant of Corynebacterium glutamicum 9366-AEC/100/Leu-6, resistant to AEC and dependent on homoserine and leucine. The strain produced 47.0g, 21.5g and 50.8g L-lysine/L of medium when cultured in medium containing sucrose, ethanol, and acetic acid respectively as carbon source.

Sato et al. (1990) studied production of L-lysine by a mutant strain of *Brevibacterium flavum* MJ 233-L-11 resistant to AEC. The strain produced 4.4g L-lysine/L of medium in 24 hours when cultured in a glucose based medium (9g/L) supplemented with yeast extract, biotin, thiamine HCl, and inorganic salts. Wibowo et al. (1992) isolated a mutant strain of *Brevibacterium flavum* UCL 260 resistant to AEC, capable of producing L-lysine on molasses medium. The strain was develop from AEC resistant strain ATCC 21475. The experimental results showed that fermentation with fed batch system produced higher levels of L-lysine concentrations (Y p/s = 0.31) at a comparable growth rate. Furthermore strain UCL 260 showed its stability during fermentation.

Production of L-lysine by auxotrophic homoserine or threonine dependent strain of Micrococcus glutamicus was also reported by Bucko et al. (1989). The yield of L-lysine was significantly increased by inoculation with a sample collected from pre-cultured medium at the beginning of the stationary growth phase. The L-lysine production after 76 hours was 68g/L vs. 56g/L after 92 hours in a conventional process. Yoshihara et al. (1990) studied L-lysine production by culture of phenylalanine- resistant Brevibacterium or Corynebacterium sp. A strain of *Brevibacterium lactofermentum* AJ 12435 (FERM – 2294) was treated with N-methyl-N-nitro-N-nitrosoguanidine (MNNG). The resulting mutant was cultured in a medium containing m-trifluromethyl-phenylalanine to give resistant strain AJ 12437 (FERM-BP-2297). This strain when cultured in a glucose based medium, produced 16.1 g/l of L-lysine vs 9.4 g/l, for the parent strain in 48 hours. In a similar type of study, Lin et al. (1990) reported that Corynebacterium Pekinese, AS 1563, when mutated with nitrosoguanidine the resulting mutant produced twice the L-lysine that the parent strain produced. Very interestingly, a novel mutant of Corynebacterium glutamicum was described by Won et al. (1990), producing very large amounts of L-lysine (120 g L-lysine/l culture medium). Schrumpf et al. (1992) cited the isolation of an L-lysine AEC resistant hyper-producing strain MH20-22B of Corynebacterium glutamicum, from a wild strain of Corynebacterium glutamicum. The strain produced 44 g/l of L-lysine HCl from 100g/l of glucose in a simple mineral salts medium.

Nomura et al. (1980) reported L-lysine production by *Brevibacterium flavum* QL-5, derived from Brevibacterium flavum No. 2247 in a medium containing glucose supplemented with threonine, biotin and inorganic salts. Misra et al. (1979 and 1980), reported L-lysine producing strains AECR-13 and AECR-25 of *Corynebacterium glutamicum* resistant to AEC obtained by mutation of Corynebacterium glutamicum ATCC 13059 with UV light. The L-lysine production on glucose medium under shake flasks condition was about 45 g/l after 196 hours. However, the same strain produced 45.1 g/l and 43 g/l in fermenter in 120 hours and 72 hours repectively. Another strain, AECR 25 under shake flask condition produced 40 g/l in 96 hours. However, the same strain produced 39.2 g/l L-lysine on cane molasses medium in fermenter after 96 hours. Strain improvement for L-lysine production was also reported by Guha et al. (1982). The wild strain of Corynebacterium glutamicum was treated with MNNG for isolation of AEC resistant mutants having increased L-lysine producing ability. One strain (CAEC-75), isolated from 10 mM AEC containing plate, produced 24 g L-lysine/l. Induced mutation and strain selection techniques were adopted to increase L-lysine production by strain CAEC-75. Using physical and chemical mutagens in a stepwise fashion, strain CUEU-40 was selected which was 50% more productive than CAEC-75. Breeding of drug resistant *Brevibacterium flavum* was reported by Tan et al. (1983). Brevibacterium flavum AS 1495 was treated with mutagen (MNNG) and cultured in the presence of drugs, such as AEC, sulphadiazine, erythromycin and rifampicin, to produce drug resistant and high L-lysine producing mutant strain. One such train of Brevibacterium flavum FRR 961 produced 44.5 g/l L-lysine on glucose medium with a conversion rate of 35.6%. Yinghua et al. (1992) developed a mutant of *Brevibacterium flavum* Au 112 from a parent starin of Brevibacterium flavum Au 111-2 by mutation with MNNG and α-thiazol-DL-alanine (TA). The new mutant Au 112 (a TA-resistant strain) produced 15.2g of L-lysine/l of medium with a corresponding glucose conversion rate of about 33.8% under submerged fermentation condition. Hilliger et al. (1990 and 1992) cited more efficient methods for microbial manufacture of L-lysine using cultures of *Corynebacetrium glutamicum*. In one experiment, Corynebacterium glutamicum yielded 52.4 g L-lysine/l in a medium containing beet molasses (290g/l). In another experiment, after a 65 hours culture on waste water obtained from sugar production, 82.4 g L-lysine/ kg was recorded by a strain of *Corynebacterium glutamicum* k-10 IMET 11429. Broeer et al. (1993) described a process (secretion screening procedure) for the manufacture of L-lysine with bacterial mutants having enhanced amino acid secretion activity. A *Corynebacterium glutamicum* mutant, isolated by this procedure furnished 10 μmol L-lysine/min/g dry weight at pH 7.5. After 72 hours of incubation, this mutant had an internal L–lysine concentration of 25mM and an external concentration of 208 mM, while the parent strain has a concentrations of 39 mM and 72 mM respectively. Maria and Duarte (1992) reported L-lysine production by an analogue sensitive mutant of *Corynebacterium glutamicum*. A fluroacetate / fluropyruvate sensitive mutant was derived from the parent strain of *Corynebacterium glutamicum* ATCC 21513. The mutant strain (FA/FP sensitive) produced 26.5 g/l L-lysine which was 3-fold more than the amount manufactured by the parent strain.

Plachy (1970) studied effect of medium composition on the production of L-lysine by Corynebacterium sp. 9366-H-454. The strain, when grown on medium containing 10% sucrose and 4% corn steep liquor was found to produce 15.89g L-lysine / L of the medium in 96 hours. A nutrient medium containing molasses and urea, supplemented with inorganic salts was used by Dunce et al. (1980) for the production of L-lysine by submerged fermentation of Brevibacterium sp. While, a nutrient medium containing peanut meal hydrolyzate, yeast extract, and sucrose supplemented with inorganic salts and biotin was also found efficient for the production of L-lysine using Corynebacterium glutamicum (Plachy et al., 1977). The yield of L-lysine after 96 hours was recorded as 61.5 g / L. Rutkov (1983a and 1983b) observed that during L-lysine fermentation the cell growth of *Corynebacterium glutamicum* was stimulated by the addition of enzymic hydrolyzate of casein while ammonium chloride significantly increased the production of L-lysine. It was also observed that the addition of L-threonine and L-lysine inhibited L-lysine production. Lysine concentration at > 0.23 M inhibited further L-lysine formation, while L-threonine inhibition was not that of a concurrent type. Smekal (1983) described a process by partly replacing sucrose in the fermentation medium with molasses. A mutant of Brevibacterium flavum when cultured in the medium, supplemented with biotin and inorganic salts, produced 52.80 g L-lysine / L of medium after 96 hours. Pelechova et al. (1983) used paper hydrolyzates as a sugar source for L-lysine production by *Corynebacterium glutamicum or Brevibacterium sp*. A low L-lysine yield was recorded (10 - 12 g / L) after 72 hours of fermentation. However, supplementing the hydrolyzate with sucrose increased the yield of L-lysine to 20 – 24 g / L. In the later case *Brevibacterium* produced more L-lysine than *Corynebacterium glutamicum*.

Sur et al. (1991) used a medium deficient of amino acids for the production of L-lysine through fermentation. A homoserine auxotroph of *Corynebacterium glutamicum* was treated with MNNG, and resulting mutant, also auxotroph for isoleucine was isolated. The double auxotroph was genetically stable and produced more L-lysine (56 g / L) in eight days. Smekal et al. (1988) studied L-lysine fermentation with *Corynebacterium glutamicum,* using different carbon sources including sucrose. Molasses and starch hydrolyzate. L-lysine concentration of 32-42 g / L with 26% conversion was achieved after 96 hours of cultivation. An improved method for preparing L-lysine was also reported by Yonekura et al (1988), using Corynebacterium glutamicum, which were resistant to β-napthoquinoline. One such strain, H-4412 produced 52 g / L L-lysine on molasses medium. The parent strain produced only 44 g / L L-lysine under the same conditions. Plachy and Ulbert (1987) reported application of mutants sensitive to amino acids for the preparation of L-lysine by fermentation. The mutant *Corynebacterium glutamicum* 9366- T/6, sensitive to L-threonine was isolated. The mutant was able to produce 49 g / L L-lysine in a medium containing sucrose and acid hydrolyzate of peanut meal. Plachy and Ulbert (1988) further studied the biochemical and production capabilities of *Corynebacterium glutamicum* strains. An increased L-lysine formation (60 g / L over 60 hours) was recorded by maintaining a high concentration of dissolved oxygen.

An improvement of L-lysine producing strain by mutation of regulatory gene was described by Liu (1987). A number of analogue resistant mutants were isolated and screened. The most powerful L-lysine producer so obtained increased the yield of L-lysine HCl from 30% to 40% on the basis of total sugar conversion. Hilliger et al. (1989b) proposed a complex medium for the production of L-lysine by a mutant strain of *Corynebacterium glutamicum* under vigorous aeration. The organism grown in a medium containing sucrose produced 105 g L-lysine / L of medium. Conversion efficiency was 27 g L-lysine / 100 g sucrose. In another experiment, Hilliger et al. (1991) reported the effect of pH and aeration rate. At characteristic pH and aeration rate 62.2 g L-lysine / L of medium was recorded with a strain of *Corynebacterium glutamicum.*

Effect of medium composition on L-lysine production by strain of *Corynebacterium glutamicum* ATCC 21513 was described by Sassi et al. (1990). The organism was found to produce high amounts of L-lysine, when grown in a medium containing glucose, ammonium sulphate and yeast extract. Along with L-lysine production, residual sugar, and dry cell mass were also measured as a function of fermentation time. It was observed that 1 g of cell mass produced 2.96 g of L-lysine. The conversion efficiency was 44%. Kawahara et al. (1990) reported stimulatory effect of exogenous glycine betaine on L-lysine production. It was indicated that the growth rate, sugar consumption rate, and production rate of an L-lysine producing Brevibacterium lactofermentum mutant were stimulated by the addition of exogenous glycine betaine. The influence of synthetic carbohydrates (a mixture of Linear and branched $C_3 - C_7$ carbohydrates produced by HCHO condensation). On L-lysine production was also reported (Sukharevich et al., 1992). The synthetic carbohydrates increased cell autolysis primarily because of its effect on the teichoic acids. Stikans et al. (1991) reported effect of surfactants on L-lysine production. It was observed that the production of L-lysine by Brevibacterium sp. E-531 was increased by the addition of surface active agent (perfluorocarbons) in a dose-dependent manner. At 2.5 ml surfactant / L, the yield of L-lysine increased from 19.7 to 31.5 g / L of medium. Wang et al. (1991) also examined the culture conditions for the production of L-lysine by Brevibacterium sp. P1-13 in 500 ml Hinton flask and a 14 L jar fermenter. It was suggested that for obtaining high yield of L-lysine, the ammonium sulphate concentration (2%) must be maintained throughout the fermentation. In addition, culture conditions that improve conversion yields of L-lysine from Corynebacterium or Brevibacterium were also described by Pfefferle et al. (1993). Initially the concentration of carbon source in the medium was kept slightly lower that the strain required for optimum growth. Using this method, sugar to L-lysine conversion yields of 32.3% to 40% with a L-lysine content of the dried fermentation broth of 45% to 55% was recorded. In control experiments the conversion yields were 27% to 32% and the L-lysine content of the dried fermentation broth was 30.5% to 36.3%. Coello et al. (1992) studied the effects of nutritional limitations, such as phosphate and carbon source concentration, on the production of L-lysine by Corynebacterium glutamicum in continuous culture. Phosphate limited cultures at low growth rates were found favorable to L-lysine production. L-lysine was produced when a culture at low dilution rate (0.03 g / h) was maintained. It was further reported that a dilution rate of about 0.04 g / h should be maintained to assure good productivity and a L-lysine yield of 0.53 g / g. Production of L-lysine by Brevibacterium sp. P1-13 in fed-batch fermentation using molasses medium was investigated by Wang et al. (1993a). A 7.3% of L-lysine HCl was obtained with a medium containing 9% molasses (as glucose), 3% ammonium sulphate, 0.2% sodium acetate, 0.01% potassium dihydrogen phosphate, 0.5% fish soluble, and 4.5% hydrolyzate of plant protein. The feeding medium was composed of 30 to 37% molasses (as glucose), 2% ammonium sulphate, 0.2% sodium acetate, 0.01% potassium dihydrogen phosphate, 4.5% hydrolyzate of plant protein and 0.5% fish soluble, which was added during logarithmic phase at a constant rate for about 50 hours. By substitution of raw sugar for molasses in the feeding medium, the L-lysine HCl accumulated increased to 8.63% with a conversion efficiency of 46% based on the total sugar. The effects of compositions of raw sugar medium on L-lysine production by Brevibacterium sp. P1-13 were also studied by Wang et al. (1993b). Using Hinton flask culture, the addition of hydrolyzate of plant protein (HPP) enhanced both L-lysine production as well as cell growth remarkably, and showed no feedback inhibition. The strain P1-13 produced 5.1% L-lysine HCl from 13% initial sugar fermentation medium containing 4% ammonium sulphate. By jar fermenter culture, the optimum concentration of biotin, yeast powder, glycine betaine and dipotassium hydrogen phosphate were at 200 µg / L, 0.5%, 10^{-2}M, and 0.1% respectively. Further, L-lysine could be stimulated by adding molasses in the raw sugar medium. In fed-batch fermentation a 9.3% L-lysine HCl was accumulated within 60 hours. The conversion efficiency was 53% based on total sugar. In order to improve the L-lysine productivity of Brevibacterium sp. P1-13 further in raw sugar medium, Wang et al. (1994) investigated the optimum conditions of fermentation. The results revealed that increasing the dipotassium hydrogen phosphate and decreasing the ammonium sulphate concentrations during seed culture higher L-lysine yield could be achieved by strain P1-13. Udeh and Achremowich (1993) applied the response surface method and experimental central composite design for optimal medium composition for L-lysine bioprocess optimization. The factors selected for the test were glucose, ammonium sulphate, yeast extract, manganese chloride and biotin. Experiments within the optimum range gave a yield of 56.2 g / L of L-lysine. L-lysine production in continuous culture of an L-lysine hyper producing mutant of *Corynebacterium glutamicum* was reported by Hirao et al. (1989). The maximum values of L-lysine HCl concentration and volumetric productivity were recorded as 105 g / L and 5.6 g / h, respectively. Continuous fermentation processes for the production of amino acids are reported more difficult in terms of maintainenence than other continuous fermentation due to their susceptibility to contamination as well as degeneration and mutation issues. Continuous L-lysine fermentation with free cells of Corynebacterium glutamicum ATCC 21492 in stirred tank reactor was designed by Pham et al. (1990). For this purpose static and rotating ceramic membrane filtration units were used. Without the ceramic membrane, batch fermentation using same organism exhibited a volumetric productivity of 0.18 g L-lysine / L / h with a maximum L-lysine fermentation at a dilution rate of 0.96 / h, increased volumetric productivity of 1.83 g L-lysine / L / h with a maximum L-lysine concentration of 30.5 g / L was achieved. Production of L-lysine using immobilized living *Corynebacterium sp.* Cells in alginate get beads was described by Nasri et al. (1989). The immobilized Corynebacterium sp. Cells exhibited a slightly greater L-lysine production than free cells and accumulated 60 g / L of L-lysine at a maximum when cultured for 120 hours in a medium containing glucose (200 g / L) as carbon source. Production of L-lysine by free PVA-cryogel immobilized *Corynebacterium glutamicum* cells was described by Velizarov et al. (1992). Non-growing free cells as well as immobilized ones produced higher levels of L-lysine than growing free cells, when cultured 24 hours in a medium containing 80 g glucose / L as carbon source. For the improvement of L-lysine productivity, development of the continuous fermentation system by a bioreactor assembly was also attempted (lee and Cho, 1994). Primarily, optimal conditions on the whole cell immobilization of Corynebacterium glutamicum ATCC 21514 were studied and 76.2% immobilization ratio was obtained when the cells were entrapped with 4% k-carrageenan showing 4.0 kg get strength. Experimental results obtained from 14 days continuous fermentation showed 36.7% of sugar conversion to L-lysine. The productivity of L-lysine was disclosed as 4.96 mg / ml / mg dry cell weight / h which is 2.5 times and 4.1 times higher than those of the batch-wise fermentation by the intact cells and by the immobilized cells, respectively.

Broeer and Kraemer (1991) identified a specific secretion carrier system of L-lysine excretion by *Corynebacterium glutamicum*. The organism effectively excreted L-lysine when the internal L-lysine concentration was elevated. L-lysine effluex was also investigated using selected mutants which were not to regulate L-lysine biosynthesis by feedback inhibition. Secretion of L-lysine was not the consequence of unspecific permeability of the plasma membrane but was mediated by a secretion carrier which was specific for L-lysine. The L-lysine export was characterized by high activation energy and follows Michaelis-Menten type kinetics with an internal K_m of 20 mM and a V_{max} of 12 nmol / min / mg dry cells. Thus excretion can proceed against a preexisting chemical gradient and against the electrical potential, which rules out a previously suggested pore model. L-lysine excretion can also be observed in the wild-type strain especially under conditions of peptide uptake. Broeer et al. (1993) further reported that strains of *Corynebacterium glutamicum* with different L-lysine productivities may have different L-lysine excretion systems. The L-lysine excretion system of three different strains of Corynebacterium glutamicum were characterized in intact cells. Two strains (DG 52-5 and MH 20-22B) were L-lysine producers of different efficiency. They were bred by classical mutagenesis and have a feed-back resistant aspartate kinase. The 3rd strain (KK 25) was constructed from the wild type by introducing the feed-back resistant aspartate kinase gene of strain MH 20-22B into its genome. All three strains had different excretion system. Export in strain KK 25 was much slower than in the two mutants. The differences between the two strains were more subtle. The K_m and V_{max} were similar, but pH dependence and membrane potential dependence revealed differences in the intrinsic properties of the carrier system. The identification of a lysine-specific RNA element, named the *LYS* elemant, in the regulatory regions of bacterial genes involved in biosynthesis and transport of lysine was also reported (Rodionov et al., 2003). A part from *Corynebacterium glutamicum* and *Brevibacterium flavum*, a number of other auxotrophic mutants resistant to AEC have also been reported to produce L-lysine. Production of L-lysine by a methylotrophic *Bacillus sp.* was reported (Schendel et al., 1990). The strain was a homoserine auxotroph and resistant to AEC. When cultured on methanol based medium supplemented with biotin and vitamin B_{12}, produced 20 g L-lysine / L of medium. L-lysine production by auxotrophic mutants of Arthrobacter globiformis was reported by Samanta et al. (1988). By mutagenesis with N-methyl-N-nitro-N-nitrosoguanidine (MNNG), in two steps, a number of methionine plus threonine double auxotrophs were isolated. One strain MT 35 yielded 28.0 g L-lysine / L of medium in flask culture on glucose medium supplemented with biotin and inorganic salts.

Production of L-lysine by a double auxotrophic and AEC resistant mutant of *Bacillus megaterium* was reported by Chatterjee et al. (1990). The organism produced 26 g L-lysine / L of medium when cultured on a 4% glucose medium supplemented with inorganic salts (ammonium sulphate 50 mM) and biotin (1 μg / L). Screening of UV-irradiated and S-2-aminoethyl-L-cysteine resistant mutants of *Bacillus megaterium* for improved L-lysine accumulation was also reported (Obeta and Ekwealor, 2006). In a similar type of study Samanta and Bhattacharyya (1991) reported mutant strains of Arthrobacter globiformis resistant to AEC, capable of producing L-lysine. Under optimum conditions, on glucose medium, supplemented with biotin and inorganic salts, Arthrobacter globiformis produced 36 g L-lysine of medium in flask culture. In an another study, Sen (1991) also reported isolation of auxotrophic mutant strains of Arthrobacter globiformis resistant to AEC. Out of 200 mutants resistant to AEC, some produced L-lysine well above the wild strain. Production of L-lysine with mutant strains of Oerskovia turbata was reported by Hilliger et al. (1989a). Depending upon the particular strain and culture conditions 8.3 to 14.3 g L-lysine / L culture medium was produced. Manufacture of L-lysine with analogue resistant strain of Corynebacterium thermoaminogenes was also reported (Murakami et al. 1991). When the strain, resistant to AEC was grown on molasses medium supplemented with inorganic salts, produced 28 to 32 g L-lysine / L of medium in 72 hours. Seto and Harada (1969) studied the formation of L-lysine from acetic acid by homoserine auxotrophs of *Corynebacterium acetophilum* A51. The auxotrophs were obtained by treatment with M-methyl-N-nitro-N-nitrosoguanidine. One of them produced large amount of L-lysine from acetic acid. Addition of 0.4% yeast extract was most effective for L-lysine formation, giving a yield of 27 g L-lysine HCl / L of broth.

4. PATHWAY AND REGULATION OF L-LYSINE BIOSYNTHESIS:

Two independent pathways have been reported for the biosynthesis of L-lysine (Vogel, 1963). It was found that yeast, fungi, and some phycomycetes (Chitridales, Blastoladiales, and Mucorales) and euglenids (Flagellated algae) synthesize L-lysine from 2-oxaloglutarate and acetyl-CoA via the α – aminoadipate (AAA) pathway. On the other hand bacteria, actinomyces, certain lower fungi, protozoa and green plants synthesize L-lysine via diaminopimelate (DAP) pathway (Nishiyama and Nishida, 2000; Zabriskie and Jackson, 2000; Velasco et al., 2002, .Wittmann and Becker, 2007 and Torruella, et al., 2009).

4.1. α - aminoadipate pathway:

The steps in the α - aminoadipate pathway is shown in Figure – 1. The steps in this pathway were established with *Neurospora crassa* (Hogg and Broquist, 1968; Maragoudakis et al., 1967), *Scacharomyces cerevisiae* (Tucci and Ceci, 1972a and 1972b; Nishiyama and Nishida, 2000; Zabriskie and Jackson, 2000 and Xu et al., 2006). *Saccharomycopsis lipolytica* (Gaillardin et al., 1979), *Candida pelliculosa* (Takenouchi et al., 1981), *Candida albicans* (Nishiyama and Nishida, 2000), *Rhodotorula glutinis* (Kurtz and Bhattacharjee, 1975) and *Penicillium chrysogenum* (Demain and Masurekar, 1974; Masurekar and Demain, 1974). The regulation of α - aminoadipate pathway is shown in Table – 1.

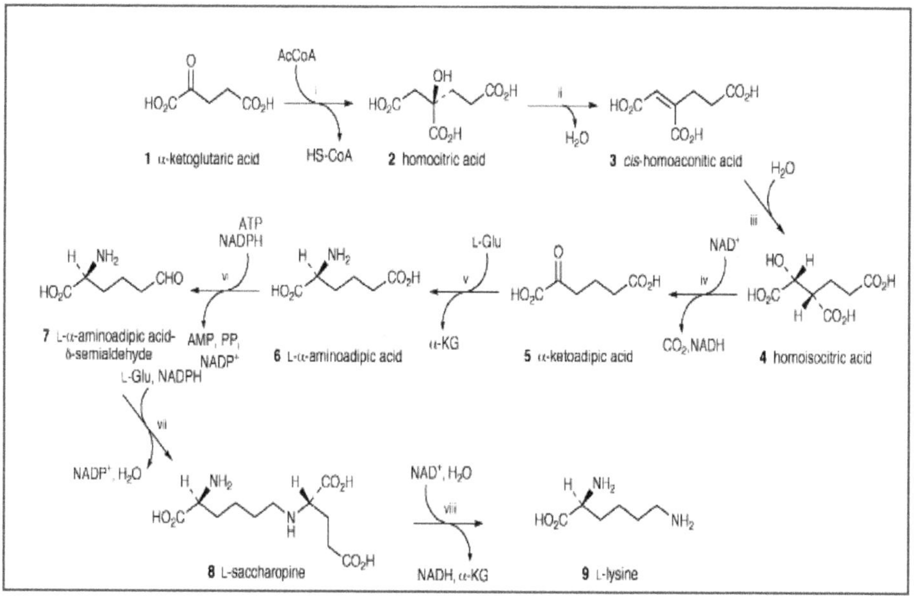

FIGURE – I: STEPS IN THE α – AMINOADIPATE PATHWAY OF L-LYSINE BIOSYNTHESIS
Enzymes of the fungal - aminoadipate pathway of lysine: i, homocitrate synthase EC 4.13.321; ii, & iii homoaconitase EC 4.2.1.36; iv, homoisocitrate dehydrogenase EC 1.1.1.87; v, aminoadipate aminotransferase EC 2.6.1.39; vi, aminoadipate reductase EC 1.2.1.31; vii scccharopine reductase EC 1.5.1.10; viii, saccharopine dehydrogenase EC 1.5.1.7.
ADAPTED FROM: Zabriskie, T.M. and Jackson, M.D. (2000). Lysine biosynthesis and metabolism in fungi. *Nat. Prod. Rep.*, **17**, 85 – 97

20

TABLE – 1: REGULATION OF THE α - AMINOADIPATE PATHWAY IN FUNGI

Enzyme	Microorganism	Gene	Reference
Homocitrate synthase-HCS (EC 2.3.3.14).	*Neurospora crassa*	-	Hogg and Broquist, 1968; Maragoudakis et al., 1967
	Saccharomyces ceravisiae *Yarrowia lipolytica*	LYS20 LYS1 -	Tucci and Ceci, 1972a and 1972b. Zabriskie and Jackson, 2000.
	Saccharomycopsisn lipolytica *Candida pelliculosa* *Rhodotorula glutinis*	- - LYS1	Gaillardin et al., 1979; Takenouchi et al., 1981; Kurtz and Bhattacharjee, 1975.
	Penicillium chrosogenum		Demain and Masurekar, 1974; Masurekar and Demain, 1974 Zabriskie and Jackson, 2000.
Homoaconitate hydratase (EC 4.2.1.36)	*Saccharomyces ceravisiae* *Aspergillus nidulans*	LYS4 / LYS7 LYSF	Bhattacharjee and Sinha, 1972 Zabriskie and Jackson, 2000. Xu, et al., 2006
Homoisocitrate dehydrogenase (EC 1.1.1.87) α-Aminoadipate aminotransferase (EC 2.6.1.39) α-Aminoadipate reductase (EC 1.2.1.31)	*Saccharomyces ceravisiae* *Saccharomyces pombe* *Candida albicans* *Penicillium chrosogenum*	LYS2 LYS1 LYS2 LYS2	Zabriskie and Jackson, 2000. Xu, et al., 2006
Scaccharopine reductase NADP⁺,L-glutamate forming (EC 1.5.1.10)	*Saccharomyces ceravisiae*	LYS9 Regulated by LYS14	Zabriskie and Jackson, 2000 Xu, et al., 2006
Scaccharopine dehydrogenase NAD⁺,L-lysine forming (EC 1.5.1.7)	*Saccharomyces ceravisiae* *Yarrowia lipolytica* *Candida albicans*	LYS1 LYS5 LYS1	Zabriskie and Jackson, 2000 Xu, et al., 2006

4.2. Diaminopimelate pathway:

The steps in the diaminopimelate pathway is shown in Figure – II and Table – II. The pathway of L-lysine biosynthesis in bacteria was first reported by Gilvarg (1958 and 1960) using a coliform bacteria representative of the principal genera of the Enterobacteriaciae. Biosynthetically L-lysine was reported to be a member of the aspartate family and the regeneration of Llysine synthesis has a close relationship with that of the other amino acids in the aspartate family, such as methionine, isoleucine and threonine (Nakayama, 1985). The diaminopimelate (DAP), L-lysine precursor was first identified in bacterial hydrolyzate (Work, 1950 and 1951) and was reported to be required for the call wall synthesis in bacteria but not for protein synthesis. It can therefore be considered another member of the aspartate family (Nakayama, 1985). Further research in this field, revealed that in bacteria, L-lysine may be synthesized from aspartate through any of the variants (four in number) of the diaminopimelate pathway. The four pathways identified as, a). succinylase pathway, b). acetylase pathway, c). dehydrogenase pathway and d). aminotransferase pathway (Tryfona and Bustard, 2005, Wittmann and Becker, 2007 and Anastassiadis, 2007). The enzymes involved in the succinylase pathway are: 1). Tetrahydrodipicolinate succinylase (DapD); 2). N-succinyl-aminoketopimelate aminotransferase (DapC); and 3). N-succinyl-diaminopimelate desuccinylase (DapE). Enzymes of acetylase pathway are: 4). Tetrahydrodipicolinate acetylase; 5). N-acetyl-aminoketopimelate aminotransferase; and 6). N-acetyl-diaminopimelate deacetylase. The dehydrogenase pathway directly forms D,L-diaminopimelate via enzyme, 7). Diaminopimelate dehydrogenase (Ddh). The enzymes of aminotransferase pathway which form L,L-diaminopimelate, include 8). Tetrahydrodipicolinate aminotransferase and 9). Diaminopimelate epimerase (DapF). Finally, L-Lysine is formed from D,L-diaminopimelate by the enzyme 10). Diaminopimelate decarboxylase.

Most bacteria only comprise one of these pathways for L-lysine synthesis (Bartlett and White, 1985 and White, 1983). The succinylase pathway is present in both gram negative and gram positive bacteria, whereas, the acetylase pathway reported to be present in exclusively some Bacullus species (Bartlett and White, 1985 and Born and Blanchard, 1999). More than one pathways have been reported in some species of the genera Corynebacterium and in *Bacillus macerans* (Bartlett and White, 1985, Malumbres and Martin, 1996 and Schrumpf et al., 1991). In *Corynebacterium glutamicum*, two different pathways based on the availability of ammonium have been reported (Blombach et al., 2009). In both pathways, synthesis L-lysine starts from oxaloacetic acid and pyruvic acid to L-piperideine-2,6-dicarboxylate. However, the formation of D,L-diaminopimelate from tetrahydrodipicolinate will take place, either through diaminopimelate dehydrogenase, when ammonium is available in excess, or by the tetrahydrodipicolinate succinylase pathway, when

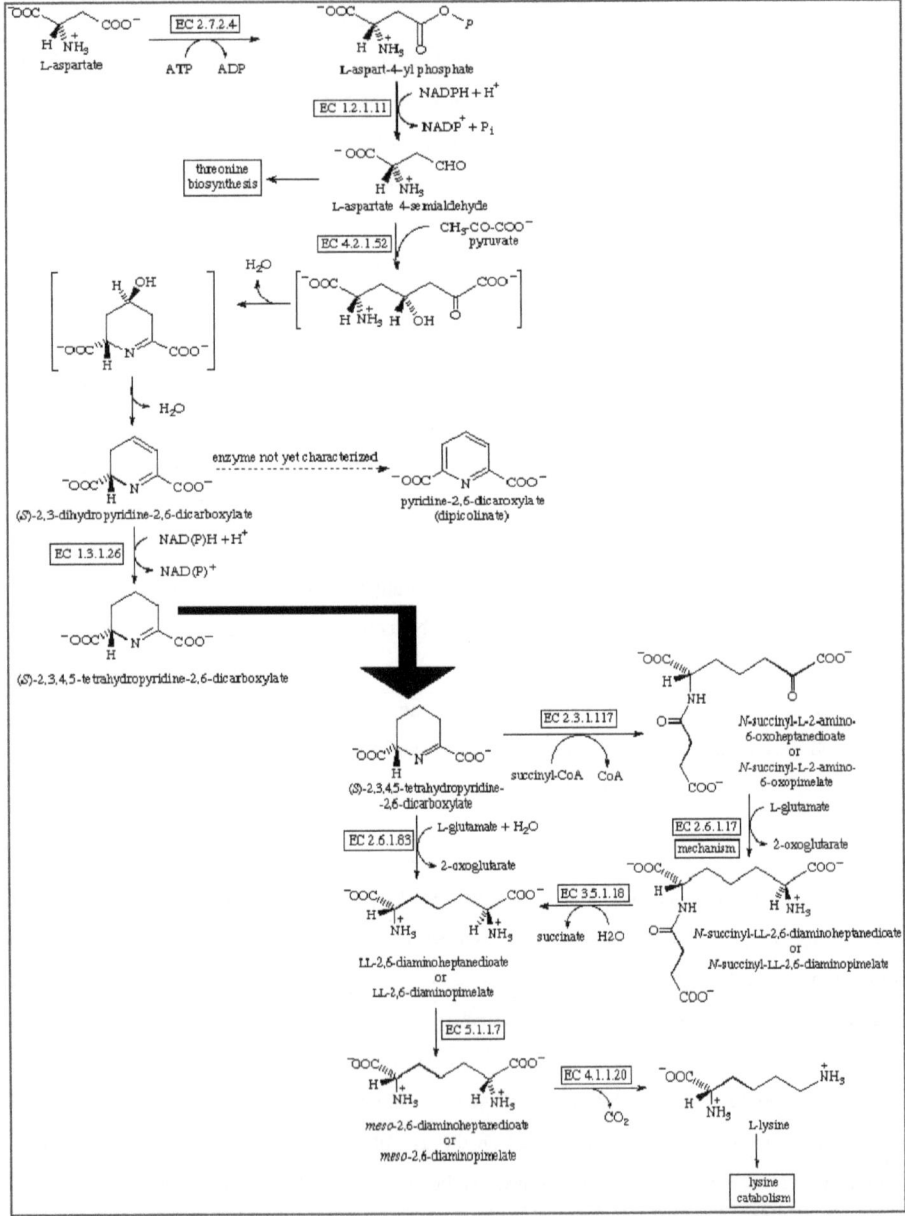

FIGURE – II : STEPS IN THE DIAMINOPIMELATE PATHWAY OF L-LYSINE BIOSYNTHESIS ADAPTED FROM: Nelson, D. L.; Cox, M. M. "Lehninger, Principles of Biochemistry" 3rd Ed. Worth Publishing: New York, 2000.

ammonium availability is low (Blombach et al., 2009). Only one isoenzyme of aspartate kinase or aspartokinase is reported to be present in *Corynebacterium glutamicum* which is encoded by two genes, *lysCα* and *lysCβ* , representing the coding sequences for the two subunits of the enzyme (Kalinowski et al., 1990, Malumbres and Martin, 1996 and Wittmann and Becker, 2007). The enzyme, aspartokinase (in wild type strains of *Corynebacterium glutamicum*) catalyzes phosphorylation of aspartate and is strongly feedback inhibited by L-lysine plus L-threonine. Thus, overexpression of the respective *LysC* gene will significantly improve the production of L-lysine (Blombach et al., 2009).

TABLE – II: REGULATION OF THE DIAMINOPIMELATE PATHWAY IN BACTERIA *(Corynebacterium glutamicum)*

Enzyme	Gene	Reference
Aspartate kinase EC 2.7.2.4	*lysC*	*Kalinowski et al., 1990* *Kalinowski et al., 2003*
Aspartae semialdehide dehydrogenase EC 1.2.1.11	*asd*	*Cremer et al., 1988* *Cremer et al.,1991*
Dihydrodipicolinate synthase EC 4.2.1.52	*dapA*	*Patek et al., 1997*
Dihydrodipicolinate reductase EC 1.3.1.26	*dapB*	*Patek et al., 1997*
Tetrahydrodipicolinate succinylase EC 2.3.1.117	*dapD*	*Wehrmann et al., 1998*
Succinyl-amino-keto-pimelate transaminase EC 2.6.1.17	*dapC*	*Hartmann et al., 2003*
Succinyl-diaminopimelate desuccinylase EC 3.5.1.18	*dapE*	*Wehrmann et al., 1998*
Meso- diaminopimelate dehydrogenase EC 1.4.1.16	*ddh*	*Cremer et al., 1998*
Diaminopimelate epimerase EC 5.1.1.7	*dapF*	*Hartmann et al , 2003*
Diaminopimelate decarboxylase E.C 4.1.1.20	*lysA*	*Cremer et al., 1998*
Lysine permease	*lysE*	*Vrljic et al., 1996* *Vrljic et al., 1999*

5. BIOMEDICAL IMPORTANCE OF L-LYSINE

L-lysine has several important biological roles (Balch and Balch, 1990; Champe and Harvey, 1993 and Rodwell, 1993, Flodin, 1997, n.d, 2010). During digestion, L-lysine bound in food protein is released as free L-lysine. There are several possible metabolic fate of the free L-lysine (Figure – III).

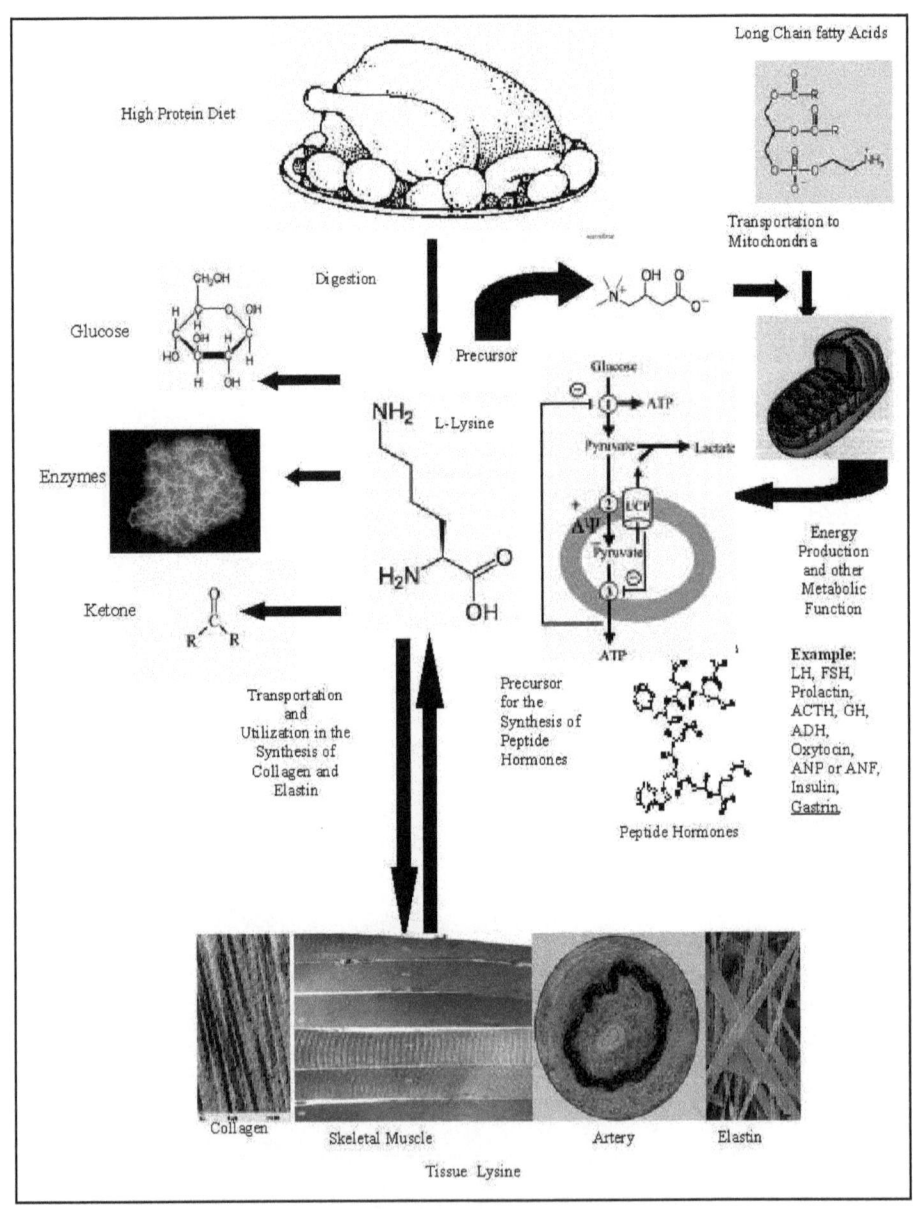

FIGURE – III: METABOLIC FATE OF FREE L-LYSINE

Some of the salient features of L-lysine are stated as under.

1. The most important one is to furnish the indispensable amino acid for synthesis of protein, particularly skeletal muscles proteins and enzymes.

2. L-lysine is one of the few amino acid that are both glycogenic and ketogenic (Figure – III). Thus L-lysine can be metabolized to yield glucose or ketone bodies when there is a deficiency of available carbohydrate and aid in the formation of D-glucose, glycogen and lipids.

3. L-lysine is an important energy source during periods of food deprivation or in some types of diabetes.

4. L-lysine is also a precursor of carnitine which is important for metabolism of fatty acids. Carnitine is an essential constituent of an enzyme associated with mitochondrial membranes. This enzyme allows long-chain fatty acids to penetrate the mitochondrial membrane and oxidized for producing energy.

5. Excess L-lysine can be utilized as source of energy as indicated by its ME value of 4600 kcal/kg. The non-nitrogenous portion of L-lysine enters the tricarboxylic acid (TCA) cycle and is eventually oxidized to carbon dioxide and water, yielding energy as ATP.

6. L-lysine plays important role in the formation of histone. When combined with nucleic acids, histone forms nucleohistone, the principal constituent of chromatin network. L-lysine also plays an important role in the formation of desmosine and isodesmosine, found in elastin, an essential constituent of yellow elastic connective tissues.

5.1. Therapeutic uses

Following are the important reported therapeutic uses of L-lysine (Griffith et al., 1978, Griffith et al., 1987, Chaitow, 1985; Balch and Balch, 1990, Pauling, 1991, McBeath and Pauling, 1993, Pauling, 1993, n.d., 1996, Flodin, 1997, n.d., 2007, and n.d, 2010).

1. As an essential building block for all protein, L-lysine is needed for proper growth and bone development in children.

2. L-lysine helps calcium absorption and maintains nitrogen balance in adults. Some researchers reported that Lysine may help prevent bone loss associated with osteoporosis.

3. Among the many functions of L-lysine is its ability to fight cold sores and herpes viruses, to aid in the production of antibodies, hormones, and enzymes, and to help in collagen formation and in the repair of tissue.

4. Because it helps to build muscle protein, it is especially important for those recovering from surgery and sports injuries.

5. L-lysine also lowers high serum triglycerides and LDL cholesterol. Thus along with vitamin C, alleviates the symptoms associated with angina pectoris.

5.2. Deficiency symptoms:

The important deficiency symptoms reported for L-lysine, are as follows:

Loss of energy, inability to concentrate, irritability, bloodshot eyes, hair loss, anaemia, retarded growth and reproduction.

5.3. Therapeutic dosages:

The reported therapeutic dosages of L-lysine for the maintenance of anti-herpes effect are 500mg to 1500mg daily, and up to 3000 mg in active stages always in divided dosage with low arginine diet (Chaitow, 1985).

5.4 Daily requirements:

The daily requirements of L-lysine for human body is as follows (West et al., 1966, Meredith et al., 1986, and Duncan et al., 1996).

Estimated minimal requirement:

		(mg / day)
1.	Infant	103
2.	Children	1600
3.	Adult male	800 (Ranges from 800 to 3,000 mg/day)
4.	Adult female	500

In general, L-lysine supplementation is very safe. Doses up to 3g daily are typically well-tolerated. Very high doses (> 10 to 15 g daily) may cause gastrointestinal upset, including nausea, abdominal cramps, and diarrhea (n.d, 2010). L-lysine supplementation is contraindicated in individuals with hyperlysinemia / hyperlysinuria, a rare genetic disorders (n.d, 2007).

6. PHARMACOKINETICS

The preferred route for L-lysine supplementation is oral. After oral administration, L-lysine is absorbed through small intestine and via active transport enters into enterocytes and then to liver through portal circulation where it is utilized with other amino acids in the synthesis of protein.It is metabolised in the liver by condensation reaction to ketoglutarate to form saccharopine. It then and converted to L-alpha-aminoadipic acid semialdehyde and finally become acetoacetyl-CoA (Flodin, 1997). In human, absorption rates are similar to those from digestion of proteins, suggesting supplementation is an effective and efficient means of correcting a ditery L-lysine deficiency (Flodin, 1993, n.d, 2007).

7. MECHANISM OF ACTION

The mechanism of action of L-lysine is based on it's conversion into acetyl CoA through a complex mechanism and the production of energy (Figure – IIIA). L-lysine is also reported (Broquist, 1982) as precursor of the carnitine, responsible for transporting long-chain fatty acids into the mitochondria for energy production and other metabolic function. Once L-lysine is bound to a polypeptide structure, biosynthesis of carnitine is initiated by methylation of one of lysine's amino groups. Transformation of this same amine group is also involved in the biosynthesis of collagen and elastin (n.d, 2007)

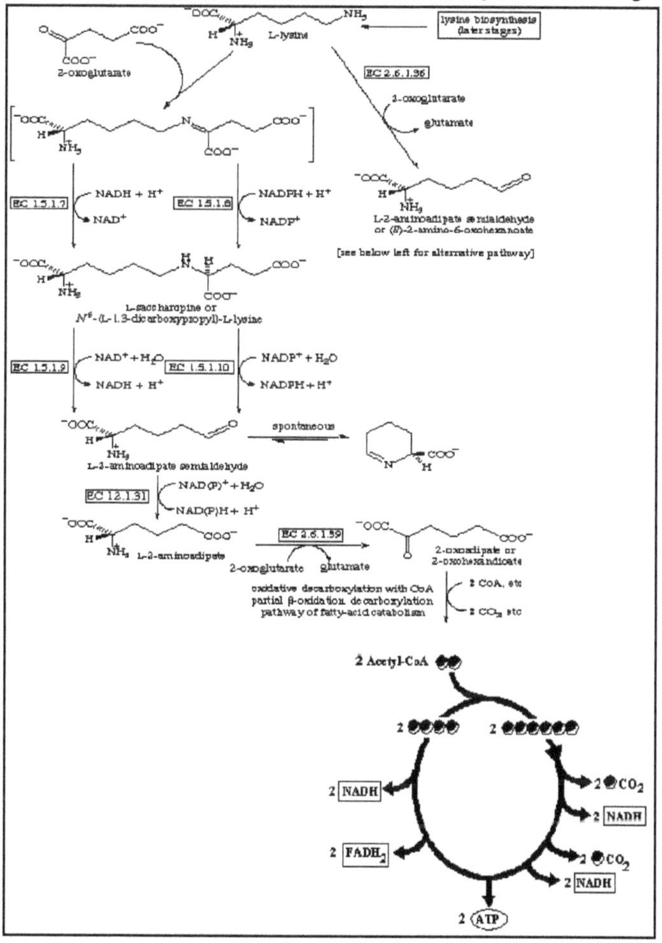

EC 1.2.1.31 L-aminoadipate-semialdehyde dehydrogenase
EC 1.5.1.7 saccharopine dehydrogenase
EC 1.5.1.8 saccharopine dehydrogenase
EC 1.5.1.9 saccharopine dehydrogenase
EC 1.5.1.10 saccharopine dehydrogenase
EC 2.6.1.36 L-lysine 6-transaminase
EC 2.6.1.39 2-aminoadipate transaminase

Energy Generation Details :
6 NADH's generated ,
2 FADH$_2$ is generated ,
2 ATP are gener ated,
4 CO$_2$'s are released
MODIFIED FROM:
Nelson, D. L. and
Cox, M. M. (2000)
Lehninger, Principles of
Biochemistry, 3rd Ed.
Worth Publishing:
New York, pp.153-6

FIGURE – IIIA: MECHANISM OF CONVERSION OF L-LYSINE INTO ACETYL CoA FOR THE PRODUCTION OF ENERGY

28

EXPERIMENTAL PROCEDURES

1. ISOLATION AND IDENTIFICATION OF CORYNEBACTERIUM GLUTAMICUM FOR THE DEVELOPMENT OF AUXOTROPHS CAPABLE OF PRODUCING L-LYSINE:

1.1. PRELIMINARY SCREENING TEST:

1.1.1. Collection of soil and sewage samples:

30 sewage and 30 soil samples were collected from different parts of Karachi city.

1.1.2. Isolation medium:

A modified Bouillon medium of following composition was used for isolation of *Corynebacterium glutamicum* (Kinoshita et al., 1957a). Quantities in g/l solution used were:

Peptone	10.0 g
Meat extract	5.0 g
Yeast extract	2.0 g
Sodium chloride	2.5 g
Agar	20.0 g

Ingredients were mixed in 1000 ml. distilled water and heated to dissolve completely. Cyclohexamide at a concentration of 50 mg/l was added to restrict the growth of fungi (Daoust, 1976). The pH of the medium was adjusted to 7.0 by 0.1 N sodium hydroxide. Aliquots (10.0 ml) of this solution were decanted into MacCartney bottles and sterilized by autoclaving for 15 minutes at 15 psi.

1.1.3. Isolation of bacteria:

1.0 g of each sewage and soil samples were added to 100 ml sterile distilled water and a number of tenfold dilutions were prepared in the same diluent. Aliquots 0.1 ml of each 1:1000 diluted sewage and soil suspensions were added separately to the agar plates prepared from the isolation medium and distributed evenly over the surface with the aid of sterile glass spreading rod. Following incubation at 30°C for 48 to 72 hours. Those plates which contained sufficient number (30 to 50) of discrete, well isolated colonies were

29

selected as the source of culture to be evaluated for the production of L-glutamic acid (Daoust, 1976).

1.1.3.1. Screening medium:

A medium of following composition was used for screening test (Daoust, 1976). Quantities in g/l of the solution used were:

Glucose	25.0
Yeast extract	0.5
Meat extract	0.5
Ammonium sulphate	7.0
Di-potassium hydrogen phosphate	7.0
Potassium di-hydrogen phosphate	3.0
Agar	20.0

Salts (quantities in mg/l)

Magnesium sulphate 7 H_2O	0.52
Ferrous sulphate 7 H_2O	0.014
Manganese sulphate 4 H_2O	0.005
Sodium chloride	0.010

Growth factor (quantity in μg/l)

Biotin	1.0

Ingredients were mixed in 1000 ml. distilled water and heated to dissolve completely. The pH was adjusted to 7.2 with 0.1 N sodium hydroxide. Aliquots 10.0 ml of this solution were decanted into MacCartney bottles and sterilized by autoclaving for 15 minutes at 15 psi.

1.1.3.2. Screening test:

Replica-plating technique was used for the transfer of colonies from the isolation medium to screening medium (Lederberg and Lederberg, 1952). Using a cylinder (of the same diameter as the petri plate) covered with a sterile valveteen material with a thick nap, the well-isolated colonies growing on plates containing isolation medium were transferred by carefully pressing imprints (replicas) from the cylinder on to fresh plates containing screening medium in duplicate. The position of colonies on the original and the replicated plates were carefully marked for the selection of corresponding colonies. The replicated cultures were incubated at 30°C for 48 to 72 hours to allow ample time for growth of organisms and production and diffusion of the amino acid in to the medium. At the end of incubation period, the colonies developed on screening medium were exposed to a dose of ultraviolet light from Philips germicidal tube light, 19 W kept at a distance of 52 cm for 7 minutes to kill the cells and to prevent over growth during the assay.

1.1.3.3. Detection of L-glutamic acid:

Detection of L-glutamic acid produced on screening medium by individual colonies was done by "Bio-autographic Technique" (Daoust, 1976). The agar plates were overlaid with approximately 10.0 ml. of molten L-glutamic acid assay medium seeded with test organism *Pediococcus acidilactici* ATCC-8042. Following incubation at 37°C for 24 hours, the test organism grows in the immediate area surrounding any colony that produced L-glutamic acid.

1.2.   FINAL SCREENING TEST:

1.2.1. Purification and maintenance of culture:

Organisms showing positive L-glutamic acid production by bio-autographic technique were further purified by repeated cultivation and maintained at 4°C on agar slants prepared by modified Bouillon medium (Kinoshita et al., 1957a).

1.2.2. Production medium:

The composition and pH of the production medium was similar to that of the screening medium (Daoust, 1976) with the exception of agar. Aliquots 15.0 ml of this solution were decanted in 50 ml Erlenmeyer flasks and sterilized by autoclaving for 15 minutes at 15 psi.

1.2.3. Cultivation:

Each flask was inoculated with two loopfuls of organisms from each slant culture and incubated with shaking in a Gallenkamp orbital shaker-incubator at 200 rpm. maintained at 30°C for 72 hours (Kinoshita et al., 1957a).

1.2.4. Identification of L-glutamic acid and other amino acids:

Routine identification of the L-glutamic acid and some other amino acids produced was made with a slight modification in the method used by Kinoshita et al. (1957a). Thin layer chromatography being carried out instead of paper chromatography. Aluminium T.L.C. plates of 0.2 mm thickness (DC-Mikrokarten SIF 20 x 20 cm. Riedel-De-Haen Aktiengesellschaft Seelze-Hannover) were used. Broths (8 to 10 µl) were spotted on to the plate and up flow development was made at 25°C using solvent system: n-butanol-acetic acid-water (4:1:1 v/v) for about 6 hours. After development, the chromatogram was treated with 0.15% ninhydrin-butanol solution. Spots of all amino acids were visualised as violet colour, only proline indicated yellow colour. The Rf-values were measured and compared with the Rf-values of authentic amino acid samples on separate chromatogram. T.L.C. of the authentic amino acids was always carried along with test samples. Solutions of amino acids were prepared according to the method of Brenner et al. (1969). Each solution contained 1 mg/ml of the authentic amino acid in deionized water containing 10% n-propanol by volume.

1.3. L-GLUTAMIC ACID PRODUCTION BY VARIOUS STRAINS:

1.3.1. Method:
The quantitative estimation of L-glutamic acid produced by various strains was carried out by "Turbidimetric Assay" (Shockman, 1963 and Barton-Wright, 1952).

1.3.2. Test organism:
Pediococcus acidilactici ATCC-8042.

1.3.3. Assay medium:
Dehydrated L-glutamic acid assay medium was used. The medium contained all growth factors and amino acids necessary for the growth of *Pediococcus acidilactici* except L-glutamic acid. To rehydrate, 105.0 g of the medium was suspended in 1000 ml. distilled water and heated to dissolve completely. 5.0 ml of the medium were added into each tube (10 x 100 mm) to be used for the development of standard curve (Calibration curve) and also to each tube containing material under assay. Sufficient distilled water was added to

make up the total volume to 10.0 ml. The tubes were autoclaved for 10 minutes at 15 psi.

1.3.4. Stock cultures of the test organism:

Stock cultures of *Pediococcus acidilactici* ATCC-8042 were prepared by stab inoculation of Bacto-Micro Assay Culture Agar. After 48 hours of incubation at 37°C, the tubes were stored in refrigerator. Subculturing were made at monthly intervals in triplicate.

1.3.5. Preparation of inoculum:

Inoculum for assay was prepared by subculturing from a stock culture to 10.0 ml. of Bacto-Micro Inoculum Broth. After 24 hours of incubation at 37°C, the cells were centrifuged and the supernatant liquid was decanted. The cells were resuspended in 10.0 ml. of sterile isotonic sodium chloride solution. The suspension was further diluted with sterile isotonic sodium chloride solution to obtain a 1:10 ml. dilution. One drop of this suspension was used to inoculate each of the assay tubes (10 ml).

1.3.6. Standard curve:

A standard curve (Figure – IV) was obtained by using L-glutamic acid at levels of 0.0, 3.0, 6.0, 9.0, 12.0, 15.0 and 18.0 ug/ml. The concentration of L-glutamic acid required for drawing the standard curve was prepared by dissolving 1.0 g L-glutamic acid in 1000 ml. sterile distilled water. This was the stock solution (1000 µg/ml.). It was further diluted in sterile distilled water to prepare the required dilution.

AOD = NOD adjusted to agree with Beer's Law

NOD = Observed optical density (OD) minus the blank

OD = Observed at wavelength 750 nm.

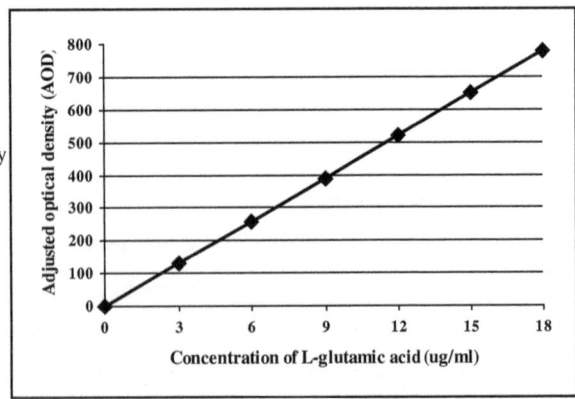

FIGURE – IV: STANDARD CURVE FOR THE DETERMINATION OF L-GLUTAMIC ACID

1.3.7. Procedure:

Each tube (standard as well as test) was inoculated with one drop (about 0.05 ml.) of the inoculum prepared from the test organism. The tubes were incubated at 37°C in a constant temperature water bath for 24 hours. At the end of incubation period all the assay tubes were removed from the water bath, dried, and shaken. The turbidity so produced was read in a spectrophotometer (Spectronic 21) at 750 nm.

1.3.8. Calculation and results:

A standard curve was first drawn by plotting the response of the test organism (AOD) against increasing concentration of L-glutamic acid. The sample (fermented broth) to be tested was appropriately diluted and the amount of L-glutamic acid was then determined with the help of standard curve (Figure-IV).

1.4. IDENTIFICATION OF CORYNEBACTERIUM GLUTAMICUM

Indentification of *Corynebacterium glutamicum* was primarily based on the taxonomic comparison. The characteristic morphological, cultural, bio-chemical, physiological and chemical properties were taken into consideration (Bergey's Manual of Systematic Bacteriology - Vol. 2 1986; Bergey's Manual of Determinative Bacteriology, 9th Ed, 1994; Goodfellow and Schaal, 1979). Details of the methods employed for morphological, cultural, bio-chemical, physiological and chemical studies are given in Annexure - II, III, IV, V and VI (A, B, and C) respectively. *Corynebacterium glutamicum* ATCC-13032 was used for taxonomic comparison.

1.4.1. Summary of the routine tests used for the identification of bacteria:

1.4.1.1. Morphological characteristics:

Morphological characteristics include the study of the cell shape, cell arrangement, Gram's staining, acid-fast staining, capsule staining, flagella staining and spore staining. Details of the methods employed are given in Annexure - II.

1.4.1.2. Cultural characteristics:

Cultural characteristics include the study of growth, surface structure, pigmentation, etc., on nutrient agar plates, nutrient agar stabs, nutrient broth and tellurite medium. Details of the methods employed are given in Annexure - III.

1.4.1.3. Bio-chemical and physiological characteristics:

The bio-chemical and physiological tests were performed according to the prescribed methods MacFaddin, 1976; and Cappuccino and Sherman 1983). Details of the methods employed for bio-chemical and physiological tests are given in Annexure - IV and V respectively.

1.4.1.4. Chemical characterictics:

Chemical characteristics include the study of principal amino acids and amino sugars (Yamada and Komagata, 1970a); mycolic acid and other long-chain components in whole-organism methanolysates (Minnikin et al., 1975) and DNA nase composition (Marmur and Doty, 1962; Deley and Schell, 1963; and Yamada and Komagata, 1970a). Details of the methods employed for chemical tests are given in Annexure - VI (A,B and C).

2. ISOLATION OF AUXOTROPHIC MUTANTS FROM PARENT STRAIN:

Isolation of auxotrophic mutants (nutritional mutants) was carried out by the modified penicillin selection technique followed by UV irradiation. The whole process is shown in Figure - V.

2.1. Organism:

Corynebacterium glutamicum FRL-44.

2.2. Culture media:

Stock cultures were maintained on agar slants of complete medium (Tatum, 1945). A minimal medium was used to grow cells for irradiation (Davis and Mingioli, 1950). Composition and quantities in g/l of complete and minimal medium were as follows;
Complete medium: (Tatum, 1945)

Meat extract	10.0
Peptone	10.0
Yeast extract	5.0
Sodium chloride	3.0
Agar	15.0

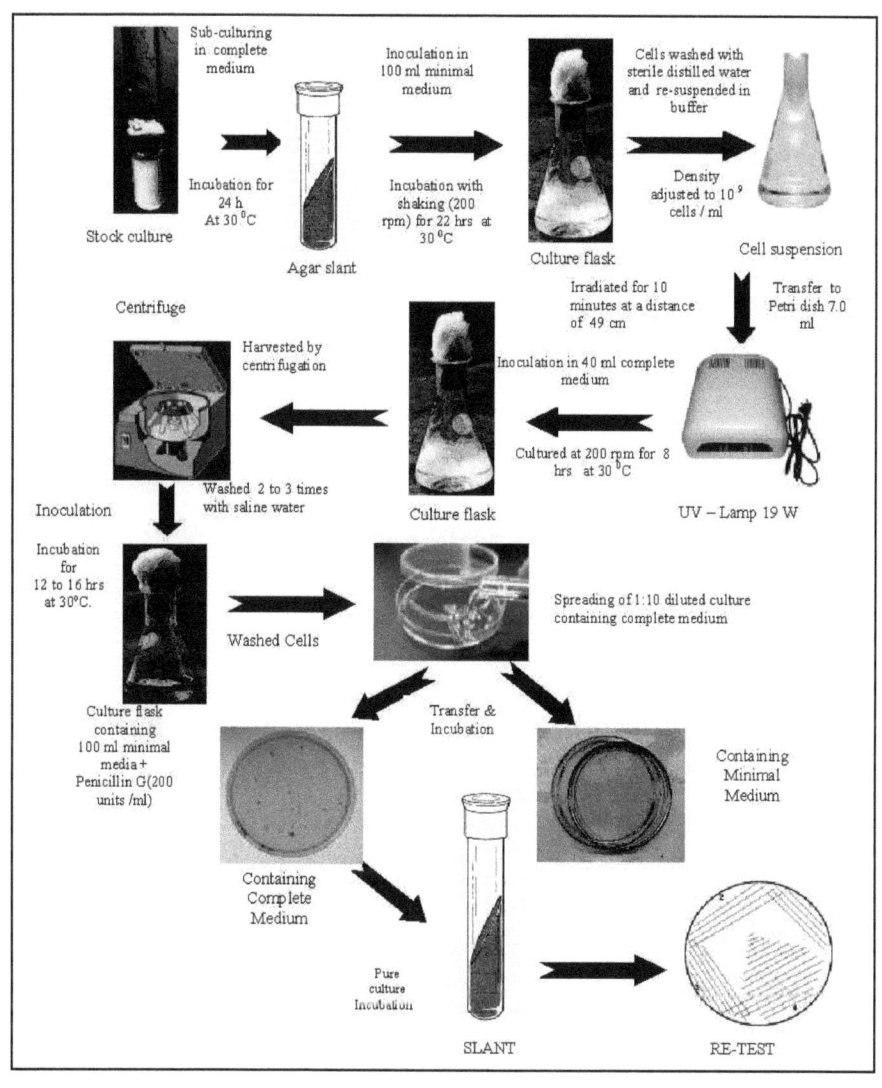

FIGURE – V: MODIFIED PENICILLIN SELECTION TECHNIQUE FOR THE ISOLATION OF AUXOTROPHIC MUTANTS

The ingredients were mixed in 1000 ml distilled water and heated to dissolve completely. The pH of the medium was adjusted to 7.2 by 0.1 N sodium hydroxide. Aliquots 10.0 ml of this solution were decanted into MacCartney bottles and sterilised by autoclaving for 15 minutes at 15 p.s.i.

Minimal medium: (Davis and Mingioli, 1950)

Glucose	2.0
Di-potassium hydrogen phosphate	7.0
Potassium di-hydrogen phosphate	3.0
Manganese sulphate $4H_2O$	0.1
Ammonium sulphate	1.0
Sodium citrate $3H_2O$	0.5
Agar	15.0

The ingredients were mixed in 1000 ml. distilled water and heated to dissolve completely. The pH of the medium was adjusted to 7.2 by 0.1 N sodium hydroxide. Aliquots 10.0 ml of this solution were decanted into MacCartney bottles and sterilised by autoclaving for 15 minutes at 15 p.s.i.

2.3. Mutagenesis (UV irradiation):

Ultraviolet irradiation was carried out with a 19 W Philips germicidal tube light mounted in Gallenkamp orbital shaker-incubator. From the overnight cultured agar slant prepared with complete medium, a loopful of the culture was transferred to 100 ml. of minimal medium in 250 ml Erlenmeyer flask in triplicate. The flasks were incubated with shaking at 200 rpm for 22 hours at 30°C. At the end of incubation period, the cells were washed with sterile distilled water and resuspended in buffer containing (7.0 g of Na_2HPO_4, 3.0 g of KH_2PO_4 and 4.0 g of NaCl/l. water) to a density of 10^9 cells/ml (approximately) using spectrophotometric and dilution-platting procedure (Cappuccino and Sherman, 1983). The tube light was allowed to warm up for a standard period of 20 minutes and then 7.0 ml. of the cell suspension was irradiated for 10 minutes in a petri dish of 9 cm. diameter at a distance of 49 cm. The viable cell counts were made in minimal medium before and after irradiation. (Adelberg and Myers, 1953; Karlstrom, 1965 and Nakayama et al., 1961a and 1961b).

2.4.   Isolation of auxotrophic mutants:

Irradiated   suspension was transferred to a 250 ml. Erlenmeyer flask   containing 40   ml. complete medium. It was cultured at 200 rpm. for 8 hours at 30°C. Cells were harvested by centrifuging, washed 2 to 3 times with saline water, and then inoculated to minimal medium containing penicillin G at a concentration of 200 units/ml. After incubation in the medium for 12 to 16 hours, cells were harvested by centrifuging and spread out onto the complete agar medium with 1:10 dilution. Colonies which appeared after incubation for 1 to 2 days on this agar were tested for their growth on the complete and minimal medium. Each colony which showed no growth on the minimal medium was transferred to a slant and its growth on the minimal, and complete medium was tested again by inoculation from this slant. Colonies were tested by regular streaking on agar plate under laminar flow hood (Adelberg and Myers, 1953 and Nakayama at al., 1961a and 1961b).

.

2.5.   Identification of required substance:

The   identification   of growth factor requirements was   made   auxanographically (Pontecorvo, 1949) using following substance.

Amino acids: L-lysine, L-methionine, L-homoserine, L-isoluecine, L-threonine, L-aspartate and L-glutamate.

Vitamins: Thiamine, riboflavin, pantothenic acid, nicotinic acid, pyridoxal   and cyanocobalamin.

Petri dishes (20 cm. dia.) containing 20 ml. of minimal agar medium were inoculated with 1.0 ml. (containing about 10^8 cells) of the strain to be tested. Penassay filter paper discs (3 mm. in dia) were soaked in growth factor solutions containing 1 mg/ml of standard material. The discs were then placed on the surface of the medium and incubated at 37°C for 24 to 48 hours. The substance promoting the growth of the mutant showed a circular area of growth around the discs. The desired auxotrophic types were rechecked by streaking on appropriate solid medium and maintained on a monthly transfer schedule on Difco nutrient agar slants at 4°C.

2.6.   L-lysine producing ability of mutant strains:

2.6.1. Medium: (Nakayama et al. 1961a and 1961b).

A   modified medium with increased glucose and biotin concentrations and   yeast extract instead of meat extract of following composition was used. Quantities in g/l solution were as follows;

Glucose	70.0 g.
Peptone	10.0 g.
Yeast extract	5.0 g.
Sodium chloride	3.0 g.
Potassium di-hydrogen phosphate	0.5 g.
Di-potassium hydrogen phosphate	1.5 g.
Magnesium sulphate	0.5 g.
Ammonium sulphate	5.0 g.

Biotin	25.0 µg/l

The ingredients were mixed in 1000 ml distilled water and heated to dissolve completely. The pH of the medium was adjusted to 7.2 with 0.1 N sodium hydroxide. Aliquots (30.0 ml) of this solution were decanted into 250 ml Erlenmeyer flasks and sterilised by autoclaving for 15 minutes at 15 psi.

2.6.2. Cultivation:

Each flask was inoculated with a loopful of respective bacterial culture grown on nutrient agar slants for 48 hours. After shaking at 200 rpm. at 30°C for 96 hours, cells were harvested by centrifugation and the clear liquid was subjected for qualitative and quantitative estimation of L-lysine.

2.6.3. Qualitative analysis:

L-lysine and other accumulated amino acids were identified with authentic samples on T.L.C plate. The mehtod and solvent system were same as described in section 1.2.4. (Identification of L-glutamic acid other amino acids).

2.6.4. Quantitative analysis:

The quantitative estimation of L-lysine was carried out by "Turbidimetric Assay" (Shoekman, 1963 and n.d.,1953).

2.6.4.1. Test organism:

Pediococcus acidilactici ATCC-8042

2.6.4.2. Assay medium:

Dehydrated L-lysine assay medium was used. The medium contained all growth factors and amino acids necessary for the growth of *Pediococcus acidilactici* except L-lysine. To rehydrate, 105 g of the medium was suspended in 1000 ml. distilled water and heated to dissolve completely. 5.0 ml. of the medium was added to each tube (10 X 100 mm) to be used for the standard curve and also to each tube containing material under assay. Sufficient distilled water was added to make up the total volume to 10.0 ml. The tubes were autoclaved for 10 minutes at 15 psi.

2.6.4.3. Stock cultures of the test organism:

Stock cultures of *Pediococcus acidilactici* ATCC-8042 were prepared by stab inoculation in Bacto-Micro Assay Culture Agar. After 48 hours of incubation at 37°C, the tubes were stored in refrigerator. Subculturing were made at monthly intervals, in triplicate.

2.6.4.4. Preparation of inoculum:

Inoculum for assay was prepared by subculturing from a stock culture to 10.0 ml. of Bacto-Micro Inoculum Broth. After 24 hours of incubation at 37°C, the cells were centrifuged and the supernatant liquid was decanted. The cells were resuspended in 10.0 ml. of sterile isotonic sodium chloride solution. The suspension was further diluted with sterile isotonic sodium chloride solution to obtain a 1:10 ml. dilution. One drop of this later suspension was used to inoculate each of the assay tubes (10 ml).

2.6.4.5. Standard curve:

A standard curve was obtained by using L-lysine at levels of 0.0, 3.0, 6.0, 9.0, 12.0, 15.0, 18.0, 21.0 and 24.0 µg/ml. The concentration of L-lysine required for drawing the standard curve was prepared by dissolving 1.0 g L-lysine in 1000 ml. sterile distilled water. This was the stock solution (1000 µg/ml.). It was stable for 2 months when stored at 2°C to 6°C. under toluene. The stock solution was further diluted in sterile distilled water to prepare the required dilution.

2.6.4.6. Procedure:

Each tube (standard as well as test) was inoculated with one drop (0.05 ml.) of the inoculum prepared from the test organism. The tubes were incubated at 37°C in a constant temperature water bath for 24 hours. At the end of incubation period all the assay tubes were removed from the water bath, dried, and shaken. The turbidity thus produced was measured in a spectrophotometer (Spectronic 21) at 750 nm.

2.6.4.7. Calculation and results:

A standard curve was first drawn by plotting the response of the test organism (AOD) against increasing concentration of L-lysine. The sample (fermented broth) to be tested was appropriately diluted and the amount of L-lysine was then determined with the help of standard curve (Figure - VI).

AOD = NOD adjusted to agree
with Beer's Law
NOD = Observed optical density
(OD) minus the blank
OD = Observed at wavelength
750 nm.

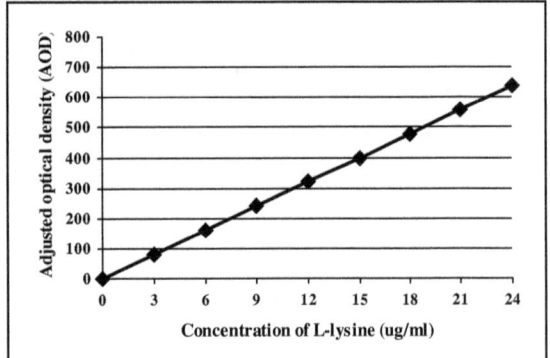

FIGURE – VI: STANDARD CURVE FOR THE DETERMINATION OF L-LYSINE

3. ISOLATION OF S-(2-AMINOETHYL)-L-CYSTEINE (AEC)-RESISTANT MUTANTS:

3.1. Organism:

Corynebacterium glutamicum FRL-2753 a homoserine auxotroph (derived from *Corynebacterium glutamicum* FRL-44) was used as a parent strain. Organisms were maintained on a monthly transfer schedule on nutrient agar slants at 4°C.

3.2. Culture medium:

Organisms from the stock culture were transferred to a complete medium of Tatum (1945). The minimal medium of Davis and Mingioli (1950) was used to grow cells after treating with the mutagen. The composition of complete and minimal medium was same as described for the isolation of auxotrophic mutants from parent strain (Sectionn 2.2).

3.3. Mutagen:

N-methyl-N-nitro-N-nitrosoguanidine (MNNG) was used as mutagen.

3.4. Isolation of AEC - resistant mutants: (Sano and Shiio, 1970).

Over night cultured cells in the complete medium were diluted to 1:40 with fresh medium and incubated for 5 hours at 30°C to obtain exponentially growing cells. The cells were treated with 2 mg/ml of N-methyl-N-nitro-N-nitrosoguanidine (MNNG) in 0.1 M phosphate buffer for 30 minutes at 0°C. The cells were then washed with sterile saline solution and inoculated directly into minimal medium supplemented with S-(2-Aminoethyl)-L-Cysteine (AEC) 2.0 mg/ml, L-threonine, 2.0 mg/ml and DL-methionine 50 µg/ml. Colonies that appeared on the surface of agar plate during 2 to 7 days of incubation were picked up as AEC - resistant mutants (Sano and Shiio, 1970).

3.5. L-lysine producing ability of AEC-resistant strains:

Materials and methods used were same as described for L-lysine producing ability of mutant strains (Section - 2.6)

4. OPTIMIZATION OF L-LYSINE PRODUCTION FROM CORYNEBACTERIUM GLUTAMICUM (FRL No.3960).

4.1. Effect of different concentrations of glucose (carbon source) and ammonium sulphate (nitrogen source) on L-lysine production:

The materials and methods were similar to those described for L-lysine producing ability of auxotrophic mutants (Section - 2.6). The dry cell weight was determined by the method of Kim et al. (1981); while the sugar (residual) was calculated by the method of Morris (1948), modified by Neish (1952). Different concentrations of glucose (70, 90, 110, 130, 150 and 170 g/l) and ammonium sulphate (5, 10, 15, 20, 25 and 30 g/l) were added to medium I, II, III, IV, V and VI respectively. The effect of these variables on the production of L-lysine was recorded. Sodium acetate (2.0 g/l) was also added in the media.

4.2. Effect of biotin on L-lysine production:

Medium-III, containing 110 g/l glucose with variable concentration of biotin (20 µg/l to 200 µg/l) was used in the study. The materials and methods were same as described for the L-lysine producing ability of auxotrophic mutants (Section - 2.6).

4.3. EFFECT OF VITAMIN B₁ ON L-LYSINE PRODUCTION:

Medium -III containing 110g/l glucose and 125µg/l biotin was used with varying concentration of Vitamin B_1 (from 0.1 mg/l to 2.5 mg/l). The materials and methods were same as described for the L-lysine producing ability of auxotrophic mutants (Section - 2.6).

4.4. EFFECT OF AERATION ON L-LYSINE PRODUCTION:

With the exception of cultivation technique, rest of materials and methods were similar to those described for L-lysine producing ability of auxotrophic mutants (section - 2.6). The cultivation was carried out using 25 ml to 200 ml of medium - III (containing 110 g/l glucose, 125 µg/l biotin and 1.0 mg/l Vitamin B_1) in 500 ml Erlenmeyer flasks.

4.5. DETERMINATION OF DRY CELL WEIGHT: (Kim et al., 1981)

The cell paste was obtained from the culture broth by centrifugation, and dried in a oven at 105°C until constant cell weight was obtained.

4.6. ESTIMATION OF SUGAR:

Reagants:

(a)   Anthrone reagent: 2.0 g anthrone was dissolved in 1.0 l of 95 % sulphuric acid.

(b)   Sulphuric acid 95%: Prepared by the addition of 1.0 l of concentrated
         sulphuric acid to 50.0 ml. of distilled water.

(c) Standard glucose solution: Prepared by the addition of 500 mg of anhydrous D-glucose in distilled water and the volume was adjusted to exactly 250.0 ml(Morris, 1948, modified by Neish, 1952)

4.6.1. Standard curve:

Anhydrous glucose solutions containing 50 to 300 µg/ml in distilled water were transferred to Spectrophotometric tubes, and to this 8.0 ml of the anthrone reagent was added to the tubes. The tubes were left at room temperature for 15 minutes. Absorbence was measured at 540 nm against blank prepared with 1.0 ml of distilled water instead of the sample. The standard curve was prepared by plotting absorbance against glucose concentration (Figure-VII).

43

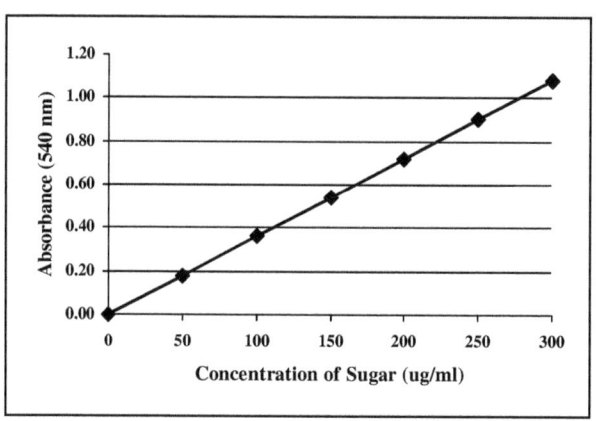

FIGURE – VII: STANDARD CURVE FOR THE DETERMINATION OF SUGAR

4.6.2. Procedure:

Samples of the fermented broth to be tested were appropriately diluted and the reaction was carried out in the same way as described earlier for the standard curve. The concentration of sugar in the samples were determined with help of standard curve.

5. PRODUCTION OF L-LYSINE ON GLUCOSE AND MOLASSES MEDIUM IN MINI JAR FERMENTER.

5.1. Organism:

Corynebacterium glutamicum FRL No. 3960, a homoserine auxotroph and resistant to AEC was used. Organism was maintained on a monthly transfer schedule on nutrient agar slants.

5.2. Seed medium:

A medium of following composition was designed as seed medium. Quantities in g/l solution were as follows:

Glucose	30.0 g.
Peptone	10.0 g.
Yeast extract	5.0 g.
Sodium chloride	3.0 g.
Potassium di-hydrogen phosphate	0.5 g.

Di-potassium hydrogen phosphate		1.5 g.
Magnesium sulphate		0.5 g.
Ammonium sulphate		0.5 g.
Sodium acetate		0.1 g.

Biotin		25.5 µg/l

The ingredients were mixed in 1000 ml. distilled water and heated to dissolve completely. The pH of the medium was adjusted to 7.2 with 0.1 N sodium hydroxide. Aliquots (30.0 ml) of this solution were decanted into 250 ml. Erlenmeyer flasks and sterilised by autoclaving for 15 minutes at 15 psi.

5.3.   Fermentation medium:

Two types of media of following composition were designed and used as fermentation medium. Quantities in g/l solution were as follows:

	Medium I	Medium II
	(Glucose medium)	(Molasses medium)
Glucose	110.0 g.	----
Molasses	--	400.0 g.
Yeast extract	5.0g	---
Soybean protein acid hydrolyzate	---	20.0 g.
Peptone	10.0 g.	---
Cornsteep liquor	---	5.0 g.
Fish meal	---	5.0 g.
Meat extract	5.0 g.	---
Potassium di-hydrogen phosphate	0.5 g.	0.5 g.
Di-potassium hydrogen phosphate	1.5 g.	1.5 g.
Ammonium sulphate	15.0 g.	15.0 g.
Calcium carbonate	10.0 g.	10.0 g.
Magnesium sulphate 7 H_2O	0.25 g.	0.25 g.
Manganese sulphate 4 H_2O	0.01 g.	0.01 g.
Ferrous sulphate 4 H_2O	0.01 g.	0.01 g.
Sodium acetate	2.00 g.	2.00 g.

| Vitamin B$_1$ | 0.001 g. | 0.001 g. |
| Biotin | 125.0 µg/l | 25.0 µg/l |

The ingredients were mixed in 1000 ml. distilled water and heated to dissolve completely. The pH of the medium was adjusted to 7.2 with 0.1 N sodium hydroxide. To this 1.0 ml. of 20% silicon RD antifoam in water was added and the whole medium was transferred to 2.0 l capacity minijar fermenter (Eyla- Tokyo Rikakikai Co. Ltd. Japan), marked as No. 1 (containing glucose medium) and No: 2 (containing molasses medium) and sterilized by autoclaving for 15 minutes at 15 psi.

5.4.   Cultivation:

Stage-I: Each flask (in triplicate) containing 30.0 ml. seed medium was inoculated with a loopful of culture grown on nutrient agar slant for 24 hours. The flasks were shaken at 200 rpm at 30°C for 24 hours in a Gallenkamp orbital shaker-incubator to prepare seed culture.

Stage-II : The minijar fermenter (No:1) containing 1.0 l glucose medium and fermenter (No:2) containing molasses medium were inoculated with the contents of one seed flask. The fermenter were stirred at 400 rpm with an air flow of 1.0 l/l of medium/minute at 30°C. The cultivation was terminated after 96 hours. When required 0.1 ml of silicon antifoam was added to the medium.

5.5.   Separation and purification of L-lysine:

Ion exchange method for the separation and purification of L-lysine (Samejima, 1972) was modified and used. At the end of cultivation, the temperature in the fermenter was raised to 60°C for 15 minutes. The whole culture liquor was then transferred to a 2.0 l Erlenmeyer flask. The broth was centrifuged at 10,000 rpm in a Gallenkamp high speed centrifuge for 20 minutes to remove the cells and other precipitated material.

The supernatant was adjusted to pH 2.0 with 6 N HCl and passed through a column (30 cm in length and 6 cm in diameter) packed with 250 g (about 20 cm in the column) of strong cation exchange resin-Amberlite IR -120 (TM of Rohm and Haas Co.) The supernatant was passed with a flow rate of 100 drops/ minute to adsorb L-lysine. After washing the column twice with 250 ml of deionized water, the column was eluted with 500 ml of dilute aqueous ammonia (prepared by diluting 375 ml. of strong ammonia solution to 1000 ml. with distilled water). After washing the column twice with 250 ml. of deionized water the whole liquid was passed with the same flow rate through second column of same dimension, packed with 250 g of a weak cation exchange resin - Amberlite IRC - 50 (TM of Rohm and Haas Co.)

After washing the column with 250 ml. of deionized water, the whole solution (about 1.0 l) was adjusted to pH 5.5 with 6 N HCl. The solution was treated with activated carbon (10 g/l). After 6 hours, the carbon-lysine slurry was filtered by applying vacuum and concentrated to about 30% under reduced pressure and kept at 10°C for over night. The material was filtered, and the crystalline material (L-lysine hydrochloride) thus obtained was washed with cold deionized water and dried in a vacuum oven.

5.6   Test for purity:

5.6.1 Thin layer chromatography

Thin-layer chromatographic method was used to detect the purity of the L-lysine (The Pharmacopoeia of Japan, 11th Ed. English version, 1986). 0.10 g of sample was dissolve in 10.0 ml. of deionzed water and used as the sample solution. Standard and test sample solutions were (5.0 µl of each) spotted on Aluminium T.L.C. plate of 0.2 mm thickness (DC-Mikrokarten SIF-10 X 20 cm. Riedel-De-Hean Aktiengesellschaft Seelze-Hannover), and developed with a mixture of n-propanol and strong ammonia water (67:33) to the distance of about 10 cm and dried in air. After spraying a solution of ninhydrin in acetone (0.5 g of ninhydrin in 1000 ml acetone), the plate was dried at 80°C for 5 minutes to observe the spots. Purity of the sample was detected by the presence of only one purple spot of same Rf-value as that of standard L-lysine.

5.6.2. Infrared absorption spectrum:

The infrared absorption spectrum was obtained with Jasco IR A-1 Grating Spectrophotometer and analysed as per standard procedure (Pavia et al., 1979). Purity of the sample was detected by the presence of similar maxima at the same wavelength as that of Standard L-lysine (The U.S.P 26; NF 21, 2003). The spectral data of the L-lysine obtained from different sources and the authentic sample is given as under:

L-lysine (Authentic sample - Mori Shita Jintan, Japan):

IR Λ max (KBr) cm^{-1}: 2910, 2100, 1620, 1500, 1410, 1360, 1340, 1290, 1188, 1160, 1000, 936, 900, 860, 738, 700. (Figure - XXIII A, Section - Discussion)

L-lysine (on glucose medium)

IR Λ max (KBr) cm^{-1}: 2900, 2080, 1610, 1495, 1404, 1345, 1320, 1290, 1187, 1135, 1000, 926, 900, 855, 731, 700, 651, 540. (Figure - XXIII B, Section - Discussion)

L-Lysine (on molasses medium)

IR Λ max (KBr) cm^{-1}: 2900, 2080, 1597, 1500, 1404, 1345, 1320, 1290, 1189, 1137, 1000, 930, 900, 860, 732, 700, 650, 543. (Figure - XXIII C, Section - Discussion)

5.6.3 Assay:

90 mg. of the sample (accurately weighed in a 125 ml flask) was dissolved in a mixture of 3.0 ml. of formic acid and 50.0 ml. of glacial acetic acid. After adding 10.0 ml. of merucric acetate solution, the flask was titrated with 0.1 N perchloric acid. The end point was determined potentiometrically. A blank determination was also performed to make any necessary correction. Each ml. of 0.1 N perchloric acid is equivalent to 9.133 mg of L-lysine hydrochloride (The U.S.P 26; N.F. 21, 2003).

RESULTS

1. ISOLATION AND IDENTIFICATION OF CORYNEBACTERIUM GLUTAMICUM FROM SEWAGE AND SOIL SAMPLES.

For the isolation of *Corynebacterium glutamicum*, sewage and soil samples (thirty each) were collected from different parts of Karachi city. Bouillon medium was used after some modifications as isolation medium. To restrict the growth of fungi, cyclohexamide at a concentration of 50 mg/l was added in the medium. In total, twentythree hundred (2300) bacterial strains, twelve hundred (1200) from sewage and eleven hundred (1100) from soil samples were isolated and tested for the accumulation of L-glutamic acid. The screening test for L-glutamic acid producing bacterial strains was carried out with the media containing carbohydrate and nitrogen source as chief ingredients supplemented with inorganic salts.

Replica-plating technique was used to transfer the organisms from isolation medium to screening medium. While, Bio-autographic technique was used for the detection of bacterial strains capable of producing L-glutamic acid using dehydrated L-glutamic acid assay medium and *Pediococcus acidilactici* ATCC-8042 as test organism. Following incubation at 37°C for 24 hours, the test organism grows in the immediate area surrounding any given colony that produced L-glutamic acid. The intensity of growth (diameter or thickness of the colony) designated as high (+ + +), medium (+ +) and low (+) which ultimately indicated the formation of L-glutamic acid by each isolate. Ninety six (96) strains out of twenty three hundred (2300) were found to produce L-glutamic acid (Table-III), seventy four (74) from sewage samples and twenty two (22) from soil samples.

All ninety six (96) strains showing positive results were then subjected to purification into the same medium to evaluate their L-glutamic acid producing capability and to identify *Corynebacterium glutamicum*. Organisms were cultivated in Erlenmeyer flask containing screening medium at 30°C for 72 hours in an orbital shaker at 200 rpm. The qualitative estimation of L-glutamic acid and any other amino acid so produced was checked by thin layer chromatography. Rf-values were measured and compared with that of authentic amino acids. Table-IV shows the Rf-values obtained from the authentic amino acids, while Table-V shows the list of amino acids identified from broth of all ninety six (96) strains after comparing the Rf-values with that of authentic amino acids.

TABLE – III: SCREENING OF BACTERIAL STRAINS AS L-GLUTAMIC ACID PRODUCERS

SERIAL No.	STRAIN (FRL No.)	SOURCE	SAMPLE No	INTENSITY OF GROWTH RECORDED	PLACE OF COLLECTION KARACHI
1	15	Sewage	1	+	Landhi
2	23	Sewage	1	+ +	Landhi
3	31	Sewage	1	+	Landhi
4	44	Sewage	2	+ + +	Nazimabad
5	54	Sewage	2	+ +	Nazimabad
6	63	Sewage	2	+ +	Nazimabad
7	85	Sewage	3	+ + +	S.I.T.E
8	98	Sewage	3	+	S.I.T.E
9	106	Sewage	3	+ +	S.I.T.E
10	131	Sewage	3	+	S.I.T.E
11	142	Sewage	5	+ + +	Gulberg
12	159	Sewage	5	+	Gulberg
13	170	Sewage	5	+ + +	Gulberg
14	182	Sewage	5	+ +	Gulberg
15	198	Sewage	6	+ +	Karimabad
16	201	Sewage	6	+ +	Karimabad
17	215	Sewage	7	+ + +	Nasirabad
18	231	Sewage	7	+ +	Nasirabad
19	244	Sewage	8	+	Sumnabad
20	263	Sewage	9	+ + +	Azizabad
21	296	Sewage	9	+ +	Azizabad
22	310	Sewage	9	+	Azizabad
23	322	Sewage	11	+ +	Alkaram Sq.
24	345	Sewage	12	+	Dastagir
25	380	Sewage	14	+ +	Alnoor Sty.
26	393	Sewage	14	+	Alnoor Sty.
27	431	Sewage	15	+ + +	F.C. Area
28	455	Sewage	15	+ +	F.C. Area
29	462	Sewage	16	+ +	N. Nazimabad
30	470	Sewage	16	+	N. Nazimabad
31	489	Sewage	16	+	N. Nazimabad

TABLE – III (Continued)

SERIAL No.	STRAIN (FRL No.)	SOURCE	SAMPLE No	INTENSITY OF GROWTH RECORDED	PLACE OF COLLECTION KARACHI
32	493	Sewage	17	+ +	Sakhi Hassan
33	519	Sewage	17	+	Sakhi Hassan
34	601	Sewage	17	+ +	Sakhi Hassan
35	625	Sewage	17	+ + +	Sakhi Hassan
36	633	Sewage	19	+ +	Buffer zone
37	650	Sewage	19	+	Buffer zone
38	671	Sewage	20	+ +	Haideri
39	688	Sewage	21	+ +	Surjani Town
40	713	Sewage	21	+ + +	Surjani Town
41	740	Sewage	21	+	Surjani Town
42	754	Sewage	21	+ +	Surjani Town
43	775	Sewage	22	+	New Karachi
44	781	Sewage	22	+	New Karachi
45	788	Sewage	22	+	New Karachi
46	797	Sewage	23	+ +	Korangi
47	801	Sewage	23	+	Korangi
48	813	Sewage	23	+	Korangi
49	822	Sewage	24	+ +	Mauripur
50	835	Sewage	24	+	Mauripur
51	850	Sewage	24	+ +	Mauripur
52	866	Sewage	24	+	Mauripur
53	873	Sewage	25	+ +	Gulshan Iqbal
54	889	Sewage	25	+ + +	Gulshan Iqbal
55	901	Sewage	25	+	Gulshan Iqbal
56	915	Sewage	26	+	P.E.C.H.S
57	926	Sewage	26	+ +	P.E.C.H.S
58	931	Sewage	26	+	P.E.C.H.S
59	942	Sewage	26	+	P.E.C.H.S
60	955	Sewage	27	+ +	K.D.A. SCH. 1
61	962	Sewage	27	+ + +	K.D.A. SCH. 1
62	973	Sewage	27	+ +	K.D.A. SCH. 1

TABLE – III (Continued)

SERIAL No.	STRAIN (FRL No.)	SOURCE	SAMPLE No	INTENSITY OF GROWTH RECORDED	PLACE OF COLLECTION KARACHI
63	982	Sewage	27	+	K.D.A. SCH. 1
64	990	Sewage	27	+ +	K.D.A. SCH. 1
65	998	Sewage	28	+	Mahmoodabad
66	1013	Sewage	28	+ + +	Mahmoodabad
67	1021	Sewage	28	+ + +	Mahmoodabad
68	1055	Sewage	28	+ +	Mahmoodabad
69	1061	Sewage	29	+ +	Clifton
70	1083	Sewage	29	+ +	Clifton
71	1125	Sewage	29	+ + +	Clifton
72	1139	Sewage	30	+ +	Steel Town
73	1155	Sewage	30	+	Steel Town
74	1190	Sewage	30	+ +	Steel Town
75	1215	Soil	2	+	Nazimabad
76	1339	Soil	2	+	Nazimabad
77	1378	Soil	3	+ + +	S.I.T.E
78	1444	Soil	5	+ +	Gulberg
79	1501	Soil	6	+	Karimabad
80	1562	Soil	6	+ +	Karimabad
81	1610	Soil	8	+	Sumnabad
82	1661	Soil	10	+	Liquatabad
83	1722	Soil	12	+ +	Dastagir
84	1861	Soil	14	+	Alnoor Sty.
85	1986	Soil	15	+	F.C. Area
86	1990	Soil	17	+ +	Sakhi Hassan
87	2025	Soil	19	+ + +	Buffer zone
88	2031	Soil	20	+	Haideri
89	2065	Soil	22	+	New Karachi
90	2085	Soil	24	+	Mauripur
91	2113	Soil	25	+	Gulshan Iqbal
92	2144	Soil	26	+	P.E.C.H.S
93	2172	Soil	28	+	Mahmoodabad

TABLE – III (Continued)

SERIAL No.	STRAIN (FRL No.)	SOURCE	SAMPLE No	INTENSITY OF GROWTH RECORDED	PLACE OF COLLECTION KARACHI
94	2198	Soil	28	+	Mahmoodabad
95	2210	Soil	29	+	Clifton
96	2235	Soil	30	+	Steel Town

(FRL No.)	-	Fermentation Research Laboratory Number (Department of Pharmaceutics, University of Karachi, Karachi – 75270, Pakistan
(+ + +)	-	High L-glutamic acid producing strains.
(+ +)	-	Medium degree of L-glutamic acid producing strains.
(+)	-	Low L-glutamic acid producing strains.

TABLE – IV: Rf- VALUES OF THE STANDARD AMINO ACIDS

NUMBER	AMINO ACID	ABBREVIATION	Rf - VALUE
1	Alanine	Ala	0.45
2	Arginine	Arg	0.17
3	Aspartic Acid	Asp	0.29
4	Cystine	Cys	0.21
5	Glutamic Acid	Glu	0.49
6	Glycine	Gly	0.33
7	Histidine	His	0.13
8	Isoleucine	Ileu	0.61
9	Leucine	Leu	0.71
10	Lysine	Lys	0.05
11	Methionine	Met	0.55
12	Ornithine	Orn	0.09
13	Phenylalanine	Phe	0.66
14	Proline	Pro	0.25
15	Serine	Ser	0.37
16	Threonine	Thr	0.41
17	Tryptophan	Try	0.76
18	Tyrosine	Tyr	0.59
19	Valine	Val	0.51

Chromatographic Plates -		Aluminium T.L.C. Plates (20 x 20 cm)
Solvent System	-	n-Butanol – Acetic acid – Water (4:1:1 v/v)
Solvent Migration	-	15 cm at 25°C
Duration	-	6 hours

TABLE – V: AMINO ACIDS PRODUCTION SPECTRA OF NINETY SIX BACTERIAL STRAINS

SERIAL No	STRAIN (FRL-No)	AMINO ACIDS IDENTIFIED						
		ALA	ASP	GLU	GLY	ILEU	PRO	VAL
1	15	+	-	+	+	-	+	+
2	23	+	-	+	+	-	+	+
3	31	+	-	+	-	-	-	-
4	44	-	-	+	-	-	-	-
5	54	-	-	+	-	-	-	-
6	63	-	-	+	-	+	+	-
7	85	+	-	+	+	-	+	+
8	98	-	-	+	-	+	-	+
9	106	+	-	+	-	-	-	-
10	131	+	-	+	-	-	-	-
11	142	-	-	+	-	+	+	-
12	159	-	+	+	-	-	-	+
13	170	+	-	+	-	-	-	-
14	182	+	-	+	+	-	+	+
15	198	+	-	+	+	-	+	+
16	201	-	-	+	-	+	+	-
17	215	+	-	+	-	-	-	-
18	231	-	-	+	-	-	+	-
19	244	-	+	+	-	-	-	-
20	263	-	-	+	-	-	-	-
21	296	+	-	+	+	-	+	+
22	310	-	-	+	-	+	-	+
23	322	-	-	+	-	+	+	-
24	345	-	+	+	-	-	-	-
25	380	+	-	+	-	-	-	-
26	393	-	-	+	-	+	+	-
27	431	+	-	+	-	-	-	-
28	455	+	-	+	-	-	-	-
29	462	-	+	+	-	-	-	+
30	470	-	+	+	-	-	-	-
31	489	+	-	+	-	-	-	-
32	493	-	-	+	-	+	+	-
33	519	+	-	+	+	-	+	+
34	601	+	-	+	+	-	+	+
35	625	-	-	+	-	-	-	-
36	633	-	-	+	-	-	+	-

TABLE – V (Continued)

SERIAL	STRAIN	AMINO ACIDS IDENTIFIED						
No	(FRL-No)	ALA	ASP	GLU	GLY	ILEU	PRO	VAL
37	650	+	-	+	-	-	-	-
38	671	-	-	+	-	-	-	+
39	688	+	-	+	-	-	-	-
40	713	-	-	+	-	+	+	-
41	740	-	-	+	-	+	+	-
42	754	-	-	+	-	+	-	+
43	775	+	-	+	-	-	-	-
44	781	-	-	+	-	-	-	-
45	788	+	-	+	-	-	-	-
46	797	+	-	+	-	-	-	-
47	801	+	-	+	-	-	-	-
48	813	+	-	+	+	-	+	+
49	822	-	-	+	-	+	+	-
50	835	+	-	+	-	-	-	-
51	850	-	-	+	-	+	+	-
52	866	+	-	+	-	-	-	-
53	873	-	-	+	-	+	-	+
54	889	-	-	+	-	+	+	-
55	901	+	-	+	-	-	-	-
56	915	+	-	+	-	-	-	-
57	926	-	-	+	-	+	+	-
58	931	-	-	+	-	+	+	-
59	942	+	-	+	-	-	-	-
60	955	-	-	+	-	+	-	+
61	962	-	-	+	-	+	+	-
62	973	+	-	+	+	-	+	+
63	982	+	-	+	+	-	+	+
64	990	-	-	+	-	+	+	-
65	998	+	-	+	-	-	-	-
66	1013	+	-	+	-	-	-	-
67	1021	+	-	+	-	-	-	-
68	1055	+	-	+	-	-	-	-
69	1061	-	-	+	-	+	+	-
70	1083	-	-	+	-	+	-	+
71	1125	-	-	+	-	+	-	+
72	1139	+	-	+	+	-	+	+

TABLE – V (Continued)

SERIAL No	STRAIN (FRL-No)	AMINO ACIDS IDENTIFIED						
		ALA	ASP	GLU	GLY	ILEU	PRO	VAL
73	1155	-	-	+	-	+	+	-
74	1190	-	-	+	-	+	+	-
75	1215	+	-	+	-	-	-	-
76	1339	-	-	+	-	+	+	-
77	1378	-	+	+	-	-	-	+
78	1444	+	-	+	-	-	-	-
79	1501	-	-	+	-	+	+	-
80	1562	-	-	+	-	+	-	+
81	1610	+	-	+	+	-	+	+
82	1661	-	-	+	-	+	+	-
83	1722	+	-	+	-	-	-	-
84	1861	+	-	+	-	-	-	-
85	1986	-	+	+	-	-	-	-
86	1990	-	-	+	-	-	+	-
87	2025	+	-	+	+	-	+	+
88	2031	-	+	+	-	-	-	+
89	2065	+	-	+	-	-	-	-
90	2085	-	+	+	-	-	-	+
91	2113	-	-	+	-	+	-	+
92	2144	+	-	+	-	+	-	-
93	2172	+	-	+	-	-	-	-
94	2198	-	-	+	-	-	-	+
95	2210	-	-	+	-	+	-	+
96	2235	+	-	+	+	-	+	+

(+) - Positive test

(-) - Neagative test

ALA - Alanine; ASP - Aspartic Acid; GLY - Glycine; ILEU - Isoleucine; GLU- Glutamic Acid; PRO - Proline; VAL - Valine

Identification was based on the Rf-Value and comparision with that of standard chromatogram.

The quantitative estimation of L-glutamic acid produced by all ninety six (96) strains was determined by turbidimetric method using *Pediococcus acidilactici* ATCC-8042 as test orgamism. Table-VI shows the results of the quantitative estimation of L-glutamic acid. Ten (10) strains produced significant quantity (13.0 g/l to 21.5 g/l) of L-glutamic acid, fourty one (41) strains produced 6.5 g/l to 12.8 g/l, while fourty five (45) strains produced 3.5 g/l to 6.0 g/l of L-glutamic acid. The highest production (21.5 g/l) was noted in FRL-44, followed by FRL-263 (19.8 g/l), FRL-625 (18.6 g/l), FRL-2025 (18.0 g/l), FRL-962(17.5 g/l), FRL-86 (16.8 g/l), FRL-170 (15.5 g/l), FRL-1378 (14.4 g/l), FRL-713 (13.5 g/l) and FRL-1125 (13.0 g/l).

Results of the perliminary morphological screening test are given in Table-VII. Out of ninety six (96), twenty five (25) strains were observed to posses charcteristic properties of *Coryneform* bacteria. Table-V indicates the list of twenty five (25) strains (suspected coryneform bacteria) showing accumulation of L-glutamic acid along with other amino acids, while the Rf values of mycolic acid and other long chain components in whole-organism methanolysates of these organisms are given in Table-XXV (Section – Discussion). Out of these twenty five (25) strains, only five (5) strains were tentatively identified by taxonomic comparison as *Corynebacterium glutamicum*. These includes strain FRL No: 44, 54, 263, 625 and 781. Results of the taxonomic comparison, i.e., charcteristic morphological, cultural, bio-chemical, physiological and chemical properties of all twentyfive (25) strains are given in Tables - VIII, IX, X, XI and XII (A, B, and C) respectively. The five (5) strains identified as *Corynebacterium glutamicum* showed similar morphological, cultural, bio-chemical, physiological and chemical properties. Details of the cultural characteristics of all five (5) strains identified as *Corynebacterium glutamicum* are listed in Table-XIII while morphological characteristics is shown in figure-VIII.

2. ISOLATION OF AUXOTROPHIC AND REGULATORY MUTANTS (RESISTANT TO AEC) AND THEIR L-LYSINE PRODUCING ABILITY:

Auxotrophic mutants (nutritional mutants) were isolated from one of the glutamic acid producing strain *Corynebacterium glutamicum* FRL-No. 44 (Wild strain) and were tested for their L-lysine producing ability. One strain (FRL No: 2753), showing highest L-lysine producing ability was further improved by developing a series of mutants resistant to S-(2-aminoethyl)-L-cysteine (AEC). Isolation of auxotrophic mutants was carried out by the penicillin selection technique by UV irradiation, while the idenfication of growth factor(s) requirements was made auxanographically using different amino acids and vitamins. L-lysine accumulation was recorded as

57

highest in homoserine requiring mutants followed by methionine plus threonine mutants in medium containing glucose supplemented with inorganic salts. In total sixteen hundred (1600) auxotrophs were isolated from *Corynebacterium glutamicum* FRL No. 44 and tested for their L-lysine producing capability. Sixty six (66) auxotrophs were found to produce L-lysine. These include 34 homoserine, 12 methionine plus threonine, 10 methionine, 3 threonine, 2 leucine and 5 isoleucine auxotrophs (Table - XIV). In addition 11 homoserine auxotrops were required vitamin B_1 for their growth. No other vitamins were found as growth factor by any auxotrophs (Table-XIV).

Based on highest L-lysine producing ability, a homoserine auxotrophic mutant FRL No. 2753 (Figure - IX) requiring vitamin B_1 as growth factor producing 26.4 g/l L-lysine was selected for further improvement by developing a series of mutants resistant to S-(2-aminoethyl)-L-cysteine (AEC). The AEC resistant mutants were isolated by treating the culture with N-methyl-N-nitro-N-nitrosoguanidine (MNNG) and observing the growth on a medium supplemented with AEC and AEC plus threonine. Frequences of the appearance of AEC resistant colonies are summarized in Table - XV.

In case of MNNG-treated cells (Table - XV, Experiment - 2), frequencies of the appearance of colonies resistant to AEC plus threonine were lower than those in untreated cells (Table - XV, Experiment - 1). The frequency further decreased with increasing concentration of AEC and threonine (Table -XV, Experiment - 3). The growth of the homoserine auxotroph FRL No. 2753 was inhibited to the extent of 50% in the presence of 1.0 mg/ml of AEC and this inhibition increased with the increasing concentration of AEC. The growth inhibition was markedly strengthened by the addition of threonine (Figure XI). The data concerning L-lysine producing ability suggested that the concentration of 2 mg/ml each of AEC and L-threonine was the best combination for the isolation of L-lysine producing mutants.

Out of 85 mutants isolated and tested for L-lysine production, only 5 strains showed a significant yield of L-lysine higher than the parent strain FRL No. 2753, a homoserine auxotrophic mutant. Eight (8) strains produced slightly higher than the parent strain. The yield of 7 mutant strains were remained unchanged; 6 strains produced less than that of the parent mutant strain, while 59 strains produced slightly higher quantity of L-lysine. Table - XVI shows the L-lysine producing ability of AEC resistant mutants. The highest production of L-lysine (33.0 g/l) noted in FRL No. 3960 (Figure-X) and ultimately it was selected for further studies.

TABLE – VI: QUANTITATIVE ESTIMATION OF L-GLUTAMIC ACID

STRAIN NUMBER (FRL)	L-GLUTAMIC ACID PRODUCED (g/l)	STRAIN NUMBER (FRL)	L-GLUTAMIC ACID PRODUCED (g/l)
15	3.5	601	7.2
23	6.5	625	18.6
31	4.0	633	8.0
44	21.5	650	4.2
54	7.8	671	9.2
63	9.5	688	8.6
85	16.8	713	13.5
98	5.0	740	6.0
106	8.0	754	9.6
131	6.0	775	5.0
142	10.5	781	4.5
159	5.2	788	3.8
170	15.5	797	10.0
182	8.4	801	6.0
198	7.0	813	5.8
201	6.5	822	8.5
215	11.5	835	3.8
231	6.4	850	7.4
244	4.5	866	4.6
263	19.8	873	6.6
296	6.5	889	11.5
310	5.4	901	5.0
322	7.4	915	4.6
345	6.0	926	7.8
380	8.2	931	5.6
393	4.5	942	4.2
431	12.6	955	9.0
455	9.0	962	17.5
462	3.2	973	10.4
470	4.5	982	3.5
489	6.0	990	9.8
493	8.5	998	3.6
519	3.6	1013	12.8

TABLE – VI (Continued)

STRAIN NUMBER (FRL)	L-GLUTAMIC ACID PRODUCED (g/l)	STRAIN NUMBER (FRL)	L-GLUTAMIC ACID PRODUCED (g/l)
1021	12.0	1661	5.5
1055	9.4	1722	9.0
1061	10.2	1861	4.5
1083	8.0	1986	3.8
1125	13.0	1990	8.0
1139	6.5	2025	18.0
1155	4.6	2031	6.0
1190	8.5	2065	6.0
1215	5.0	2085	5.8
1339	6.0	2113	4.5
1378	14.4	2144	4.2
1444	9.2	2172	6.0
1501	5.4	2198	3.5
1562	10.6	2210	4.5
1610	6.0	2235	3.8

Quantitative estimation of L-Glutamic Acid produced by all ninety six strains was done by turbidimetric method using Pediococcus acidilactici ATCC-8042 as test organism

TABLE – VII: PRELIMINARY MORPHOLOGICAL SCREENING OF NINETY SIX BACTERIAL STRAINS

STRAIN NUMBER (FRL)	GRAM STRAIN	ACID-FAST STAIN	SPORE STAIN
15	Gram-Positive Rod	-	-
23	Gram-Positive Rod	-	-
31	Gram-Positive Rod	-	-
44	Gram-Positive Rod	-	-
54	Gram-Positive Rod	-	-
63	Gram-Positive Rod	-	-
85	Gram-Positive Rod	-	-
98	Gram-Positive Rod	-	-
106	Gram-Positive Cocci	-	-
131	Gram-Positive Cocci	-	-
142	Gram-Positive Rod	-	+
159	Gram-Positive Rod	-	-
170	Gram-Positive Rod	-	-
182	Gram-Positive Rod	-	-

TABLE – VII (Continued)

198	Gram-Positive Rod	-	-
201	Gram-Positive Rod	-	-
215	Gram-Positive Rod	-	+
231	Gram-Positive Rod	-	-
244	Gram-Positive Rod	-	-
263	Gram-Positive Rod	-	-
296	Gram-Positive Rod	-	-
310	Gram-Positive Rod	-	-
322	Gram-Positive Rod	-	-
345	Gram-Positive Rod	-	-
380	Gram-Positive Cocci	-	-
393	Gram-Positive Rod	-	+
431	Gram-Positive Rod	-	-
455	Gram-Positive Rod	-	-
462	Gram-Positive Rod	-	-
470	Gram-Positive Rod	-	-
489	Gram-Positive Rod	-	-
493	Gram-Positive Rod	-	-
519	Gram-Positive Rod	-	+
601	Gram-Positive Rod	-	-
625	Gram-Positive Rod	-	-
633	Gram-Positive Cocci	-	-
650	Gram-Positive Cocci	-	-
671	Gram-Positive Rod	-	-
688	Gram-Positive Rod	-	-
713	Gram-Positive Rod	-	-
740	Gram-Positive Rod	-	-
754	Gram-Positive Rod	-	-
775	Gram-Positive Rod	-	+
781	Gram-Positive Cocci	-	-
788	Gram-Positive Cocci	-	-
797	Gram-Positive Rod	-	-
801	Gram-Positive Rod	-	-
813	Gram-Positive Rod	-	-
822	Gram-Positive Rod	-	-
835	Gram-Positive Rod	-	-
850	Gram-Positive Rod	-	+
866	Gram-Positive Rod	-	-
873	Gram-Positive Rod	-	-
889	Gram-Positive Rod	-	-
901	Gram-Positive Cocci	-	-
915	Gram-Positive Cocci	-	-
926	Gram-Positive Rod	-	-

TABLE – VII (Continued)

931	Gram-Positive Rod	-	-
942	Gram-Positive Rod	-	+
955	Gram-Positive Rod	-	-
962	Gram-Positive Rod	-	-
973	Gram-Positive Rod	-	-
982	Gram-Positive Rod	-	-
990	Gram-Positive Rod	-	-
998	Gram-Positive Rod	-	-
1013	Gram-Positive Rod	-	-
1021	Gram-Positive Rod	-	-
1055	Gram-Positive Cocci	-	-
1061	Gram-Positive Rod	-	-
1083	Gram-Positive Rod	-	-
1125	Gram-Positive Rod	-	-
1139	Gram-Positive Rod	-	-
1155	Gram-Positive Rod	-	-
1190	Gram-Positive Rod	-	-
1215	Gram-Positive Cocci	-	-
1339	Gram-Positive Rod	-	-
1378	Gram-Positive Rod	-	+
1444	Gram-Positive Rod	-	-
1501	Gram-Positive Rod	-	-
1562	Gram-Positive Rod	-	-
1610	Gram-Positive Rod	-	-
1661	Gram-Positive Rod	-	-
1722	Gram-Positive Rod	-	+
1861	Gram-Positive Cocci	-	-
1986	Gram-Positive Rod	-	-
1990	Gram-Positive Rod	-	-
2025	Gram-Positive Rod	-	+
2031	Gram-Positive Rod	-	-
2065	Gram-Positive Rod	-	-
2085	Gram-Positive Rod	-	-
2113	Gram-Positive Rod	-	-
2144	Gram-Positive Rod	-	-
2172	Gram-Positive Rod	-	-
2198	Gram-Positive Rod	-	+
2210	Gram-Positive Rod	-	-
2235	Gram-Positive Rod	-	-

(+)   Positive

(-)   Negative

TABLE – VIII: MORPHOLOGICAL CHARACTERISTICS OF TWENTY FIVE SUSPECTED CORYNEFORM BACTERIA

| FRL No. | GRAM STAIN | CELL FORM | | FLAGEL -LATION | MOTILITY | METACH- ROMATIC GRANULES | PLEOMO -RPHISM | SPORE STAIN | ACID FSST STAIN |
		YOUNG CULTURES (12-24 hrs old)	OLD CULTURES (72-120 hrs old)						
15	Weakly Positive	Small irregular rods. Some cells are arranged at an angle to give V-Formation.	Entirely or largely of coccoid cells.	Absent	None Motile	-	+ +	-	-
23	Weakly Positive	Slender irregular rods. Some of the rods are arranged at an angle to each other giving V-Formation	Composed mainly of short rods and few coccoid cells.	Present	Motile	-	+	-	-
31	Weakly Positive	Small irregular rods. Some cells are arranged at an angle to give V-Formation.	Entirely or largely of coccoid cells.	Absent	None Motile	-	+ +	-	-
44	Strongly Positive	Short slightly curved rods mostly in pairs, some occurring singly.	Very short, ellipsoidal to almost coccal.	Absent	None Motile	+	-	-	-
54	Strongly Positive	Short slightly curved rods mostly in pairs, some occurring singly.	Very short, ellipsoidal to almost coccal.	Absent	None Motile	+	-	-	-
63	Weakly Positive	Slender irregular rods. Some of the rods are arranged at an angle to each other giving	Composed mainly of short rods and few coccoid cells.	Present	Motile	-	+	-	-

TABLE – VIII (Continued)

FRL No.	GRAM STAIN	CELL FORM		FLAGEL-LATION	MOTILITY	METACH-ROMATIC GRANULES	PLEOMO-RPHISM	SPORE STAIN	ACID FSST STAIN
		YOUNG CULTURES (12-24 hrs old)	OLD CULTURES (72-120 hrs old)						
		V-Formation							
85	Weakly Positive	Irregular, slender rods. Some of the cells are arranged at an angle to give V-Form	Composed largely or entirely of coccoid cells	Present	Motile	-	+ +	-	-
182	Weakly Positive	Irregular, slender rods. Some of the cells are arranged at an angle to give V-Form	Composed largely or entirely of coccoid cells	Present	Motile	-	+ +	-	-
198	Weakly Positive	Slender irregular rods. Some of the rods are arranged at an angle to each other giving V-Formation	Composed mainly of short rods and few coccoid cells.	Present	Motile	-	+	-	-
244	Weakly Positive	Small irregular rods. Some cells are arranged at an angle to give V-Formation.	Entirely or largely of coccoid cells.	Absent	None Motile	-	+ +	-	-
263	Strongly Positive	Short slightly curved rods mostly in pairs, some occurring singly.	Very short, ellipsoidal to almost coccal.	Absent	None Motile	+	-	-	-
296	Weakly Positive	Slender irregular rods. Some of the rods are arranged at an angle to each other giving V-Formation	Composed mainly of short rods and few coccoid cells.	Present	Motile	-	+	-	-

TABLE – VIII (Continued)

| FRL No. | GRAM STAIN | CELL FORM | | FLAGEL-LATION | MOTILITY | METACH-ROMATIC GRANULES | PLEOMO-RPHISM | SPORE STAIN | ACID FSST STAIN |
		YOUNG CULTURES (12-24 hrs old)	OLD CULTURES (72-120 hrs old)						
455	Weakly Positive	Irregular, slender rods. Some of the cells are arranged at an angle to give V-Form	Composed largely or entirely of coccoid cells	Present	Motile	-	+ +	-	-
601	Weakly Positive	Irregular, slender rods. Some of the cells are arranged at an angle to give V-Form	Composed largely or entirely of coccoid cells	Present	Motile	-	+ +	-	-
625	Strongly Positive	Short slightly curved rods mostly in pairs, some occurring singly.	Very short, ellipsoidal to almost coccal.	Absent	None Motile	+	-	-	-
781	Strongly Positive	Short slightly curved rods mostly in pairs, some occurring singly.	Very short, ellipsoidal to almost coccal.	Absent	None Motile	+	-	-	-
801	Weakly Positive	Small irregular rods. Some cells are arranged at an angle to give V-Formation.	Entirely or largely of coccoid cells.	Absent	None Motile	-	+ +	-	-
813	Weakly Positive	Small irregular rods. Some cells are arranged at an angle to give V-Formation.	Entirely or largely of coccoid cells.	Absent	None Motile	-	+ +	-	-

TABLE – VIII (Continued)

| FRL No. | GRAM STAIN | CELL FORM | | FLAGEL-LATION | MOTILITY | METACH-ROMATIC GRANULES | PLEOMO-RPHISM | SPORE STAIN | ACID FSST STAIN |
		YOUNG CULTURES (12-24 hrs old)	OLD CULTURES (72-120 hrs old)						
873	Weakly Positive	Slender irregular rods. Some of the rods are arranged at an angle to each other giving V-Formation	Composed mainly of short rods and few coccoid cells.	Present	Motile	-	+	-	-
982	Weakly Positive	Small irregular rods. Some cells are arranged at an angle to give V-Formation.	Entirely or largely of coccoid cells.	Absent	None Motile	-	+ +	-	-
990	Weakly Positive	Irregular, slender rods. Some of the cells are arranged at an angle to give V-Form	Composed largely or entirely of coccoid cells	Present	Motile	-	+ +	-	-
1139	Weakly Positive	Slender irregular rods. Some of the rods are arranged at an angle to each other giving V-Formation	Composed mainly of short rods and few coccoid cells.	Present	Motile	-	+	-	-
1610	Weakly Positive	Small irregular rods. Some cells are arranged at an angle to give V-Formation.	Entirely or largely of coccoid cells.	Absent	None Motile	-	+ +	-	-

TABLE – VIII (Continued)

| FRL No. | GRAM STAIN | CELL FORM | | FLAGEL -LATION | MOTILITY | METACH- ROMATIC GRANULES | PLEOMO- RPHISM | SPORE STAIN | ACID FSST STAIN |
		YOUNG CULTURES (12-24 hrs old)	OLD CULTURES (72-120 hrs old)						
1990	Weakly Positive	Irregular, slender rods. Some of the cells are arranged at an angle to give V-Form	Composed largely or entirely of coccoid cells	Present	Motile	-	+ +	-	-
2235	Weakly Positive	Small irregular rods. Some cells are arranged at an angle to give V- Formation.	Entirely or largely of coccoid cells.	Absent	None Motile	-	+ +	-	-
ATCC 13032	Strongly Positive	Short rods or ellipsoids occurring singly and in pairs with few long branching cells.	Very Short, ellipsoidal to almost coccal.	Absent	None Motile	+	-	-	-

TABLE – IX : CULTURAL CHARACTERISTICS OF TWENTY FIVE SUSPECTED CORYNEFORM BACTERIA

FRL No.	Cultural Characteristics
15	Smooth, entire, circular, dull, cream in colour
23	Small, smooth, entire, convex, opaque, circular, yellow in colour
31	Small, entire, circular, dull, cream in colour
44	Smooth, entire, circular, dull, to slightly glistening, pale yellow in colour.
54	Smooth, entire, circular, dull, to slightly glistening, pale yellow in colour.
63	Small, smooth, entire, convex, opaque, circular, yellow in colour.
85	Small, smooth, entire, convex with shiny surface, deep orange in colour.
182	Small, smooth, entire, convex with shiny surface, deep orange in colour.
198	Small, smooth, entire, convex, opaque, circular, yellow in colour.
244	Smooth, entire, circular, dull, cream kin colour.
263	Smooth, entire, circular, dull, to slightly glistening, pale yellow in colour.

TABLE – IX (Continued)

296	Small, smooth, entire, convex, opaque, circular, yellow in colour.
455	Small, smooth, entire, convex with shiny surface, deep orange in colour.
601	Small, smooth, entire, convex with shiny surface, deep orange in colour.
625	Smooth, entire, circular, dull, to slightly glistening, pale yellow in colour.
781	Smooth, entire, circular, dull, to slightly glistening, pale yellow in colour.
801	Smooth, entire, circular, dull, cream in colour
813	Smooth, entire, circular, dull, cream in colour
873	Small, smooth, entire, convex, opaque, circular, yellow in colour.
982	Smooth, entire, circular, dull, cream in colour
990	Small, smooth, entire, convex with shiny surface, deep orange in colour.
1139	Small, smooth, entire, convex, opaque, circular, yellow in colour.
1610	Smooth, entire, circular, dull, cream in colour
1990	Small, smooth, entire, convex with shiny surface, deep orange in colour.
2235	Smooth, entire, circular, dull, cream in colour

TABLE – X : BIOCHEMICAL CHARACTERISTICS OF TWENTY FIVE SUSPECTED CORYNEFORM BACTERIA

Biochemical Characteristics	15	23	31	44	54	63	85	182	198	244	263	296	455	601	625	781	801	813	873	982	990	1139	1610	ATCC 1990	ATCC 2235	ATCC 13032
A. Cleavage of Carbohydrates																										
1. Arabinose	-	+	-	-	-	+	-	-	+	-	-	+	-	-	-	-	-	-	+	-	-	+	-	-	-	-
2. Xylose	-	+	-	-	-	+	-	-	+	-	-	+	-	-	-	-	-	-	+	-	-	+	-	-	-	-
3. Rhamnose	-	+	-	-	-	+	-	-	+	-	-	+	-	+	-	-	-	-	+	-	-	+	-	-	-	-
4. Glucose	-	+	-	+	+	+	+	+	+	-	+	+	+	+	+	+	-	-	+	-	+	+	-	+	-	-
5. Fructose	-	+	-	+	+	+	+	+	+	-	+	+	+	+	+	+	-	-	+	-	+	+	-	+	-	+
6. Mannose	-	+	-	+	+	+	+	+	+	-	+	+	+	+	+	+	-	-	+	-	+	+	-	+	-	+
7. Galactose	-	+	-	+	-	+	-	-	+	-	-	+	-	-	-	-	-	-	+	-	-	+	-	-	-	-
8. Sorbose	-	-	-	-	-	-	-	-	-	-	-	-	-	-	-	-	-	-	-	-	-	-	-	-	-	-
9. Lactose	-	+	-	-	-	+	-	-	+	-	-	+	-	-	-	-	-	-	+	-	-	+	-	-	-	-
10. Maltose	-	+	-	+	+	+	+	+	+	-	+	+	+	+	+	+	-	-	+	-	+	+	-	+	-	+
11. Raffinose	-	+	-	-	-	+	-	-	+	-	-	+	-	-	-	-	-	-	+	-	-	+	-	-	-	-
12. Starch	-	+	-	-	-	+	-	-	+	-	-	+	-	-	-	-	-	-	+	-	-	+	-	-	-	-
13. Glycerol	-	-	-	-	-	-	-	-	-	-	-	-	-	-	-	-	-	-	-	-	-	-	-	-	-	-
14. Mannitol	-	-	-	-	-	-	-	-	-	-	-	-	-	-	-	-	-	-	-	-	-	-	-	-	-	-
15. Sorbitol	-	-	-	-	-	-	-	-	-	-	-	-	-	-	-	-	-	-	-	-	-	-	-	-	-	-
16. Inositol	-	-	-	-	-	-	-	-	-	-	-	-	-	-	-	-	-	-	-	-	-	-	-	+	-	-
17. Sucrose	-	+	-	+	+	+	+	+	+	-	+	+	+	+	+	+	-	-	+	-	+	+	-	+	-	+

TABLE – X: (Continued)

Biochemical Characteristics	15	23	31	44	54	63	85	182	198	244	263	296	455	601	625	781	801	813	873	982	990	1139	1610	1990	2235	ATCC 13032
A. Assimilation of Organic Acids																										
1. Acetic Acid	+	+	+	+	+	+	+	+	+	+	+	+	+	+	+	+	+	+	+	+	+	+	+	+	+	+
2. L-lactic Acid	+	+	+	+	+	+	+	+	+	+	+	+	+	+	+	+	+	+	+	+	+	+	+	+	+	+
3. D-lactic Acid	+	-	+	+	+	-	+	+	-	+	+	-	+	+	+	+	+	+	-	+	+	-	+	+	+	+
4. Citric Acid	+	-	+	+	+	-	+	+	-	+	+	-	+	+	+	+	+	+	-	+	+	-	+	+	+	+
5. Formic Acid	+	-	+	-	-	-	-	-	-	+	-	-	-	-	-	-	+	+	-	+	-	-	+	-	+	-
6. Propionic Acid	+	-	+	+	+	-	+	+	-	+	+	-	+	+	+	-	+	+	-	+	+	-	+	+	+	+
7. Butyric Acid	+	-	+	-	-	-	-	-	+	+	-	-	-	-	-	-	+	+	-	+	-	-	+	-	+	-
8. Oxalic Acid	+	-	+	-	-	-	-	-	-	+	-	-	-	-	-	-	+	+	-	+	-	-	+	-	+	-
9. Adipic Acid	-	-	-	-	-	-	+	-	-	-	-	-	-	+	-	-	-	-	-	-	-	-	-	-	-	-
10. Gluconic Acid	+	-	+	-	-	-	+	+	-	+	-	-	-	-	-	-	+	+	-	+	+	-	+	+	+	-
11. Uric Acid	+	-	+	-	-	-	-	-	-	+	-	-	-	-	-	-	+	+	-	+	-	-	+	-	+	-

70

TABLE – XI: PHYSIOLOGICAL CHARACTERISTICS OF TWENTY FIVE SUSPECTED CORYNEFORM BACTERIA

STAIN NUMBER (FRL)	PHYSIOLOGICAL CHARACTERISTICS				
	1	2	3	4	
				(5%)	(10%)
15	25-30	6.5-7.0	A	+	-
23	30-35	6.8-7.2	FA	-	-
31	25-30	6.5-7.0	A	+	-
44	28-32	6.8-7.2	A	+	+
54	28-32	6.8-7.2	A	+	+
63	30-35	6.8-7.2	FA	-	-
85	32-37	6.8-7.0	A	+	+
182	32-37	6.8-7.0	A	+	+
198	30-35	6.8-7.2	FA	-	-
244	25-30	6.5-7.0	A	+	-
263	28-32	6.8-7.2	A	+	+
296	30-35	6.8-7.2	FA	-	-
455	32-37	6.8-7.0	A	+	+
601	32-37	6.8-7.0	A	+	+
625	28-32	6.8-7.2	A	+	+
781	28-32	6.8-7.2	A	+	+
801	25-30	6.5-7.0	A	+	-
813	25-30	6.5-7.0	A	+	-
873	30-35	6.8-7.2	FA	-	-
982	25-30	6.5-7.0	A	+	-
990	32-37	6.8-7.0	A	+	+
1139	30-35	6.8-7.2	FA	-	-
1610	25-30	6.5-7.0	A	+	-
1990	32-37	6.8-7.0	A	+	+
2235	25-30	6.5-7.0	A	+	-
13032 (ATCC)	25-37	7.0-7.2	A – FA	+	+

1: Optimum Temperature; 2. Optimum pH; 3: Oxygen Requirements; 4: Growth in NaCl

* (A – Aerobes, FA – Facultative Anaerobes)

TABLE – XII: CHEMICAL CHARACTERISTICS OF TWENTY FIVE SUSPECTED CORYNEFORM BACTERIA

(A)-AMINO ACIDS AND AMINO SUGARS COMPOSITION OF CELL WALL

STRAIN FRL No.	AMNO ACIDS						AMINO SUGARS		
	Ala	Glu	Gly	Asp	Lys	Dl-Dap	Gluco	Galact	Mura
15	+ +	+ +	+ +	-	+ +	-	+	-	+
23	+ +	+ +	tr	-	+ +	-	+	-	+
31	+ +	+ +	+ +	-	+ +	-	+	-	+
44	+ +	+ +	tr	-	-	+ +	+	-	+
54	+ +	+ +	tr	-	-	+ +	+	-	+
63	+ +	+ +	tr	-	+ +	-	+	-	+
85	+ +	+ +	+	-	+ +	-	+ +	-	+
182	+ +	+ +	+	-	+ +	-	+ +	-	+
198	+ +	+ +	+	-	+ +	-	+	-	+
244	+ +	+ +	+ +	-	+ +	-	+	-	+
263	+ +	+ +	-	tr	-	+ +	+	-	+
296	+ +	+ +	+	-	+ +	-	+	-	+
455	+ +	+ +	+	-	+ +	-	+ +	-	+
601	+ +	+ +	+	-	+ +	-	+ +	-	+
625	+ +	+ +	tr	-	-	+ +	+	-	+
781	+ +	+ +	-	tr	-	+ +	+	-	+
801	+ +	+ +	+ +	-	+ +	-	+	-	+
813	+ +	+ +	+ +	-	+ +	-	+	-	+
873	+ +	+ +	tr	-	+ +	-	+	-	+
982	+ +	+ +	+ +	-	+ +	-	+	-	+
990	+ +	+ +	+	-	+ +	-	+ +	-	+
1139	+ +	+ +	tr	-	+ +	-	+	-	+
1610	+ +	+ +	+ +	-	+ +	-	+	-	+
1990	+ +	+ +	+	-	+ +	-	+ +	-	+
2235	+ +	+ +	+ +	-	+ +	-	+	-	+

Amino Acids: Ala (Alanine); Glu (Glutamate); Gly (Glycine); Asp (Aspartic Acid);
              Lys (Lysine); Dap (Diaminopimelic Acid)

Amino Sugars: Gluco (Glucosamine); Galact (Galactasomine); Mura (Muramic Acid)

The relative amount of different amino acids and amino sugars present in chromatograms were arbitrarily scored as + + (Medium), + (Low). Tr(Trace), and – (Absent).

TABLE – XII (Continued)

(B)-DNA-BASE COMPOSITION OF ISOLATED STRAINS

STRAIN FRL No.	Tm°C	GC (Content Mole %)	STRAIN FRL No.	Tm°C	GC (Content Mole %)
15	94.5	64.1	601	95.0	62.8
23	98.6	72.2	625	92.2	58.0
31	94.3	62.9	781	91.6	56.7
44	92.2	53.8	801	94.7	62.0
54	91.6	56.6	813	94.7	62.0
63	98.9	72.7	873	98.8	72.0
85	91.3	63.7	982	95.0	63.7
182	94.0	60.2	990	95.1	62.9
198	99.0	72.4	1139	99.0	72.4
244	95.2	61.5	1610	95.3	62.2
263	91.4	53.9	1990	94.0	60.2
296	98.7	71.7	2235	94.5	64.5
455	94.6	61.7			

Tm   =   Thermal denaturation temperature

GC   =   Guanine – Cytosine

TABLE – XII (Continued)

(C)-THIN LAYER CHROMATOGRAPHIC ANALYSIS OF MYCOLIC ACID AND OTHER LONG – CHAIN COMPONENTS IN WHOLE – ORGANISM METHANOLYSATES

STRAIN FRL No.	MAMEs	FAMEs	HYDROXY FAMEs	STRAIN FRL No.	MAMEs	FAMEs	HYDROXY FAMEs
15	-	+	+	601	-	+	+
23	-	+	+	625	+	+	+
31	-	+	+	781	+	+	+
44	+	+	+	801	-	+	+
54	+	+	+	813	-	+	+
63	-	+	+	873	-	+	+
85	-	+	+	982	-	+	+
182	-	+	+	990	-	+	+
198	-	+	+	1139	-	+	+
244	-	+	+	1610	-	+	+
263	+	+	+	1990	-	+	+
296	-	+	+	2235	-	+	+
455	-	+	+				

MAMEs   =   Mycolic – Acid methyl esters

FAMEs   =   Fatty-acid methyl esters

(+)   =   Present

(-)   =   Absent

TABLE – XIII: CULTURAL CHARACTERISTICS OF STRAINS IDENTIFIED AS CORYNEBACTERIUM GLUTAMICUM

MEDIA	CULTURAL CHARACTERISTICS
Nutrient Agar Plate	Moderate growth, circular, small, slightly elevated, entire, slightly glistening pale yellow
Nutrient Agar Slant	Moderate growth, filform, dull, pale yellow
Nutrient Agar Stab	Growth occurs only on the surface
Nutrient Broth	Slightly turbid, flocculent sediment, no odour
Tellurite Medium	Dark grey to black colonies

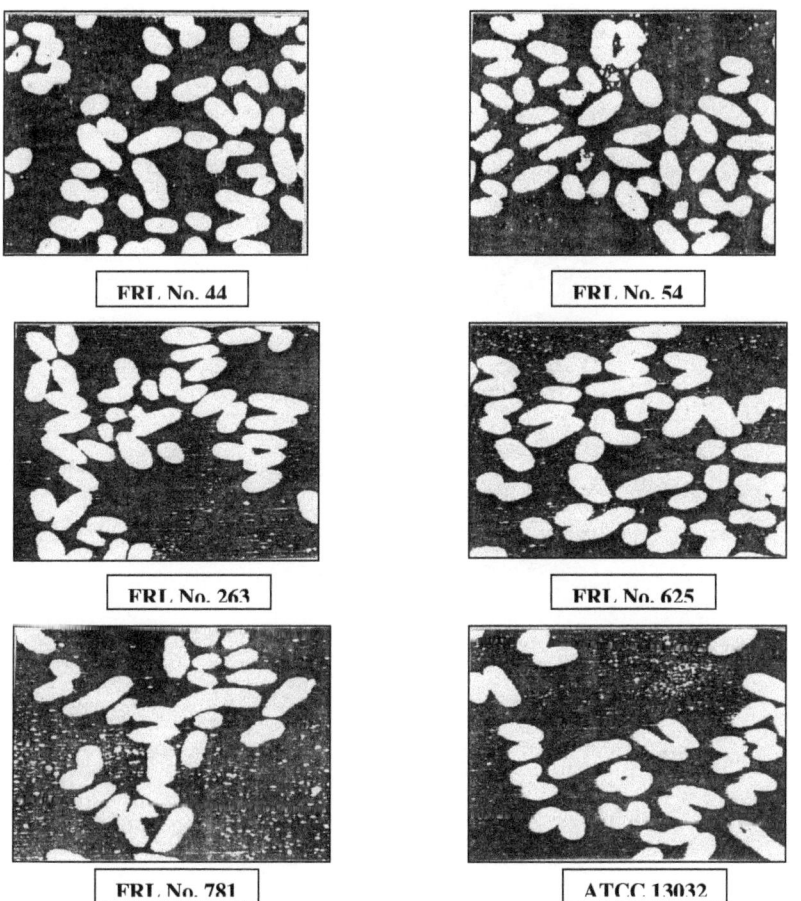

FIGURE – VIII: PHASE – CONTRAST PHOTOMICROGRAPHS SHOWING MORPHOLOGICAL CHARACTERISTICS OF CORYNEBACTERIUM GLUTAMICUM

FRL No. 44, 54, 263, 625, 781 and ATCC 13032

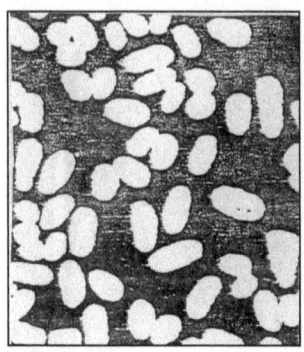

FIGURE – X: PHASE – CONTRAST PHOTOMICROGRAPH OF STRAINS FRL No. 3960

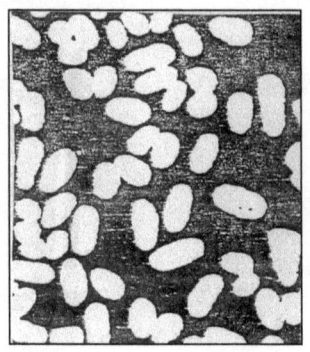

FIGURE - XI: GROWTH INHIBITION OF STRAIN FRL No. 2753 BY AEC AND AEC PLUS THREONINE

TABLE – XIV: L-LYSINE PRODUCTION BY AUXOTROPHS ISOLATED FROM CORYNEBACTERIUM GLUTAMICUM FRL No. 44

SERIAL No	STRAINS FRL No.	NUTRITIONAL REQUIREMENT (S)	GROWTH FACTOR	L-LYSINE PRODUCED (g/l)*
1	2359	Isoleucine		3.2
2	2381	Homoserine		18.6
3	2385	Homoserine		16.2
4	2409	Methionine		6.4
5	2443	Homoserine	Vit. B_1	23.5
6	2450	Isoleucine		2.8
7	2492	Homoserine		15.5
8	2517	Methionine + Threonine		12.9
9	2531	Leucine		4.0
10	2531	Methionine		4.8
11	2560	Homoserine		22.2
12	2582	Threonine		3.8
13	2589	Homoserine		21.0
14	2671	Methionine		7.2
15	2696	Isoleucine		3.0
16	2720	Homoserine		14.6
17	2737	Homoserine		18.2
18	2753	Homoserine	Vit. B_1	26.4
19	2768	Methionine + Threonine		10.2
20	2772	Homoserine		13.8
21	2793	Homoserine		24.6
22	2798	Threonine		4.0
23	2803	Homoserine	Vit. B_1	19.0
24	2868	Methionine + Threonine		12.0
25	2873	Leucine		2.6
26	2920	Homoserine	Vit. B_1	17.2
27	2931	Homoserine		18.6
28	2945	Homoserine		14.0
29	2986	Homoserine	Vit. B_1	21.8

* Values obtained after subtracting the initial concentration of L-lysine in the medium (0.4 mg/ml)

TABLE – XIV (Continued)

SERIAL No	STRAINS FRL No.	NUTRITIONAL REQUIREMENT (S)	GROWTH FACTOR	L-LYSINE PRODUCED (g/l)*
30	3017	Methionine + Threonine		10.2
31	3038	Methionine + Threonine		11.6
32	3041	Isoleucine		2.6
33	3072	Methionine		5.0
34	3079	Homoserine		16.4
35	3083	Homoserine		13.8
36	3096	Homoserine	Vit. B_1	22.0
37	3100	Homoserine		23.4
38	3166	Isoleucine		3.6
39	3192	Homoserine		13.0
40	3213	Threonine		4.8
41	3243	Homoserine	Vit. B_1	13.2
42	3266	Homoserine	Vit. B_1	25.2
43	3270	Methionine + Threonine		12.8
44	3291	Methionine		4.5
45	3350	Homoserine		14.4
46	3431	Methionine		6.2
47	3472	Methionine + Threonine		10.6
48	3490	Methionine + Threonine		11.0
49	3541	Methionine		4.6
50	3566	Methionine		5.8
51	3588	Homoserine		16.2
52	3602	Homoserine	Vit. B_1	13.5
53	3670	Methionine + Threonine		12.0
54	3681	Homoserine		17.8
55	3707	Homoserine		16.4
56	3761	Methionine + Threonine		10.2
57	3776	Homoserine		14.0
58	3780	Methionine		6.0

* Values obtained after subtracting the initial concentration of L-lysine in the medium (0.4 mg/ml)

TABLE – XIV (Continued)

SERIAL No	STRAINS FRL No.	NUTRITIONAL REQUIREMENT (S)	GROWTH FACTOR	L-LYSINE PRODUCED (g/l)*
59	3805	Homoserine		13.8
60	3810	Homoserine	Vit. B$_1$	21.0
61	3825	Homoserine		13.0
62	3831	Homoserine	Vit. B$_1$	20.6
63	3848	Methionine + Threonine		11.8
64	3868	Methionine + Threonine		10.0
65	3870	Homoserine		16.0
66	3891	Methionine		4.2

* Values obtained after subtracting the initial concentration of L-lysine in the medium (0.4 mg/ml)

TABLE – XV: FREQUENCIES OF THE APPEARANCE OF AEC RESISTANT MUTANTS FROM CORYNEBACTERIUM GLUTAMICUM FRL No. 2753

EXPERIMENT No.	MUTAGENESIS	AEC. (mg/ml)	CONCENTRATION OF L-THREONINE (mg/ml)	SURVIVING CELLS/ml
1	Non - treated	0	0	5.2×10^{-8}
		1	0	6.2×10^{-7}
		0	1	4.8×10^{-8}
		1	1	5.9×10^{-7}
2	MNNG – treated	0	0	4.9×10^{-7}
		1	0	2.4×10^{-5}
		0	1	2.7×10^{-6}
		1	1	2.2×10^{-4}
3	MNNG - treated	0	0	4.5×10^{-7}
		1	1	1.9×10^{-4}
		2	2	2.3×10^{-3}
		5	5	1.5×10^{-2}

TABLE – XVI: L-LYSINE PRODUCTION BY AEC RESISTANT MUTANTS ISOLATED FROM CORYNEBACTERIUM GLUTAMICUM FRL No. 2753

SERIAL No.	AEC RESISTANT MUTANTS FRL No.	L-LYSINE PRODUCED	L-LYSINE YIELD (%)
1	3950	28.0	6.06
2	3951	31.8	20.45
3	3952	27.6	4.54
4	3953	27.0	2.27
5	3954	26.4 (Same)	0.00
6	3955	28.6	8.33
7	3956	27.6	4.54
8	3957	31.0	17.42
9	3958	30.0	13.63
10	3959	26.8	1.51
11	3960	33.0	25.00
12	3961	25.2 (Decreased)	4.54
13	3962	28.0	6.06
14	3963	26.8	1.15
15	3964	27.2	3.03
16	3965	24.0 (Decreased)	9.09
17	3966	29.0	9.84
18	3967	26.4 (Same)	0.00
19	3968	30.2	14.39
20	3969	24.0 ((Decreased)	9.09
21	3970	27.6	4.54
22	3971	27.2	3.03
23	3972	27.0	2.27
24	3973	28.4	7.57
25	3974	32.6	23.48
26	3975	26.8	1.51
27	3976	27.0	2.27
28	3977	28.0	6.06
29	3978	28.2	6.81
30	3979	27.6	4.54
31	3980	27.6	4.54

* Corynebacterium Glutamicum FRL No. 2753 produced 26.4 g/l L-lysine

** Values obtained after subtracting the initial concentration of L-lysine in the medium of (0.4 g/l)

TABLE – XVI (Continued)

SERIAL No.	AEC RESISTANT MUTANTS FRL No.	L-LYSINE PRODUCED	L-LYSINE YIELD (%)
32	3981	26.4 (Same)	0.0
33	3982	30.0	13.63
34	3983	28.0	6.06
35	3984	27.2	3.03
36	3985	29.0	9.84
37	3986	28.0	6.06
38	3987	27.2	3.03
39	3988	28.4	7.57
40	3989	26.8	1.51
41	3990	28.4	7.57
42	3991	27.6	4.54
43	3992	28.0	6.06
44	3993	27.0	2.27
45	3994	26.4	0.0
46	3995	28.8	9.09
47	3996	27.6	4.54
48	3997	27.2	3.03
49	3998	23.8 (Decreased)	9.84
50	3999	28.0	6.06
51	4000	27.2	3.03
52	4001	26.8	1.51
53	4002	27.6	4.54
54	4003	28.2	6.81
55	4004	28.0	6.06
56	4005	27.0	2.27
57	4006	26.6	0.75
58	4007	24.0 (Decreased)	9.09
59	4008	27.2	3.03
60	4009	28.0	6.06
61	4010	27.6	4.54
62	4011	27.0	2.27

* Corynebacterium Glutamicum FRL No. 2753 produced 26.4 g/l L-lysine

** Values obtained after subtracting the initial concentration of L-lysine in the medium of (0.4 g/l)

TABLE – XVI (Continued)

SERIAL No.	AEC RESISTANT MUTANTS FRL No.	L-LYSINE PRODUCED	L-LYSINE YIELD (%)
63	4012	26.4 (Same)	0.0
64	4013	26.8	1.51
65	4014	27.6	4.54
66	4015	28.0	6.06
67	4016	26.8	1.51
68	4017	27.2	3.03
69	4018	27.6	4.54
70	4019	28.0	6.06
71	4020	27.2	3.03
72	4021	26.8	1.51
73	4022	27.6	4.54
74	4023	27.0	2.27
75	4024	26.4 (Same)	0.0
76	4025	26.8	1.51
77	4026	27.2	3.03
78	4027	27.2	3.03
79	4028	26.8	1.51
80	4029	28.0	6.06
81	4030	26.6	0.75
82	4031	27.2	3.03
83	4032	27.0	2.27
84	4033	26.4 (Same)	0.0
85	4034	27.6	4.54
86	4035	25.6 (Decreased)	3.03

* Corynebacterium Glutamicum FRL No. 2753 produced 26.4 g/l L-lysine

** Values obtained after subtracting the initial concentration of L-lysine in the medium of (0.4 g/l)

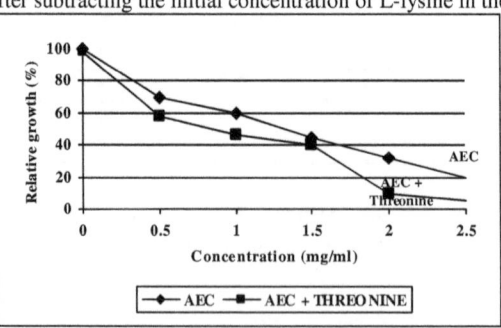

FIGURE – XI : GROWTH INHIBITION OF STRAIN FRL No. 2753 BY AEC AND AEC PLUS THREONINE

AEC = S-(2-aminoethyl)-L-cysteine

3. OPTIMIZATION OF L-LYSINE PRODUCTION FROM CORYNE BACTERIUM GLUTAMICUM FRL No. 3960

To maximize the production of L-lysine from *Corynebacterium glutamicum* (FRL No. 3960, a homoserine auxotroph-resistant to AEC), the media composition and various factors affecting the L-lysine production were studied.

Effect of glucose and ammonium sulphate:

The results presented in Table-XVII indicates that the production of L-lysine increased with the increase in glucose and ammonium sulphate concentration. The increase in L-lysine was recorded till 170 g/l glucose plus 30 g/l of ammonium sulphate added to medium - VI. The increase was also recorded below this concentration, i.e. 90 g/l, 110 g/l, 130 g/l and 150 g/l of glucose. However, at 110 g/l glucose plus 15 g/l ammonium sulphate (medium-III); negligible quantity of residual glucose was noted after 96 hours of fermentation. At this concentration 55.63 g/l and 55.64 g/l of L-lysine was produced with a mximum yield ($Y_{p/s}$) of 0.5057 g and 0.5058 g of L-lysine per unit weight of glucose (gg^{-1}) after 96 hours and 120 hours of fermentation respectively. No significant increase in the production of L-lysine and biomass was recorded by increasing the fermentation time from 96 hours to 120 hours. With the increase in glucose (i.e 130 g/l, 150 g/l and 170 g/l) plus ammonium sulphate (i.e 20 g/l, 25 g/l and 30 g/l) added to medium - IV, V and VI, a slightly higher (i.e. 57.60 g/l, 58.50 g/l and 58.66 g/l) L-lysine production was recorded after 96 hours. However, the overall yield ($Y_{p/s}$) of L-lysine, at 130 g/l, 150 g/l and 170 g/l of glucose, was significantly less (i.e. 0.443 g, 0.390 g and 0.345 g L-lysine per unit weight of glucose against respective sugar concentrations) than that of 110 g/l glucose plus 15 g/l ammonium sulphate added to medium - III (Table-XVII). Moreover, the flasks containing medium IV, V and VI also indicated significant quantities of residual glucose even after 120 hours of fermentation. In comparison, a very low L-lysine production, i.e. 32.88 g/l and 43.65 g/l with an yield ($Y_{p/s}$) of 0.4697 g g^{-1} and 0.4850 g g^{-1} were also noted in flasks containing medium-I and II respectively after 96 hours. The relationship between biomass and L-lysine production as shown in Figure-XIII, indicates production of 3.84 g L-lysine/g of biomass with medium-III after 96 hours, while 2.208g and 2.2908g of L-lysine/g of biomass was noted with medium I and II respectively. Thus flask with medium-III, containing 110 g/l glucose plus 15 g/l ammonium sulphate was noted as optimum carbon and inorganic nitrogen source level in the present study (Figure-XII).

TABLE – XVII: EFFECT OF DIFFERENT CONCENTRATIONS OF GLUCOSE AND AMMONIUM SULPHATE ON L-LYSINE PRODUCTION

Medium No.	Glucose (g/l)	Ammonium Sulphate (g/l)	Time (h)	Biomass (g/l)	Residual glucose (g/l)	L-Lysine (g/l)	Yp/s – Yield per unit of glucose (g/l)
I	70	5	36	5.90	40.50	1.55	0.0221
			48	7.00	25.35	10.90	0.2441
			72	9.86	16.05	26.66	0.3808
			96	14.89	0.35	32.88	0.4697
			120	15.00	0.34	33.02	0.4717
II	90	10	36	5.12	48.46	2.60	0.0288
			48	6.98	32.50	20.46	0.2273
			72	10.50	21.36	37.50	0.4166
			96	15.01	0.46	43.65	0.4850
			120	15.10	0.42	43.57	0.4842
III	110	15	36	6.03	60.25	6.00	0.0545
			48	8.60	30.64	22.60	0.2054
			72	10.06	10.50	40.50	0.3681
			96	14.45	0.03	55.63	0.5057
			120	15.17	0.02	55.64	0.5058
IV	130	20	36	6.00	86.55	6.12	0.0470
			48	8.50	60.36	21.29	0.1637
			72	12.44	29.45	41.50	0.3192
			96	15.25	16.66	57.60	0.4430
			120	15.60	16.85	57.90	0.4430
V	150	25	36	5.98	109.00	5.98	0.0398
			48	7.98	75.26	20.53	0.1368
			72	13.00	46.55	42.56	0.2844
			96	15.80	34.00	58.50	0.3836
			120	15.86	34.68	58.00	0.3836
VI	170	30	36	6.20	130.85	6.02	0.0354
			48	8.50	86.46	22.53	0.1325
			72	12.66	67.32	43.40	0.2252
			96	15.68	56.35	58.55	0.3385
			120	15.88	55.88	58.72	0.3401

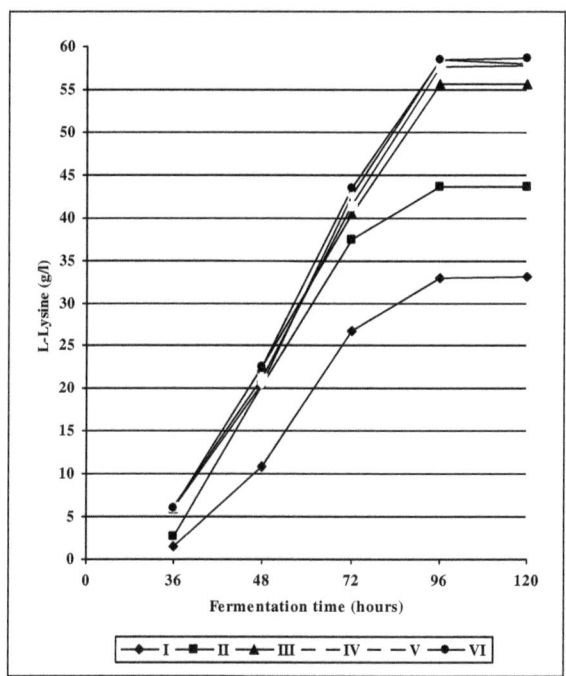

FIGURE - XII: EFFECT OF DIFFERENT CONCENTRATIONS OF GLUCOSE AND AMMONIUM SULPHATE (INCORPORTAED IN MEDIUM I TO VI) ON L-LYSINE PRODUCTION AT DIFFERENT TIME INTERVALS

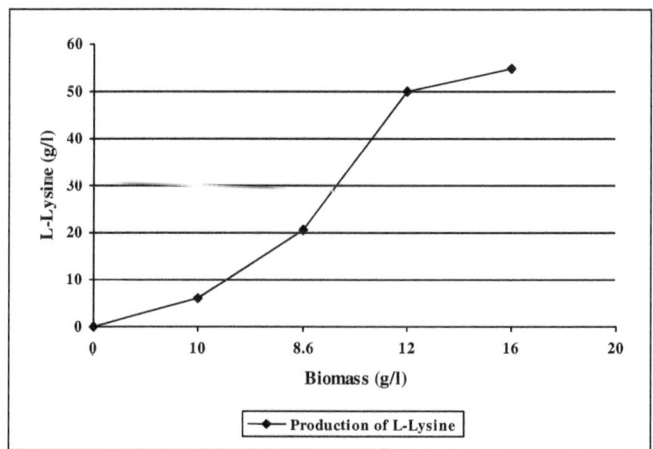

FIGURE - XIII: RELATIONSHIP BETWEEN BIOMASS AND L-LYSINE PRODUCTION USING MEDIUM – III

84

Effect of biotin:

Effect of biotin concentration in culture medium on the production of L-lysine was investigated. The relationship of L-lysine production to biotin concentration is shown in Figure-XIV. The strongest stimulatory effect on L-lysine production was achieved when 125 µg/l of biotin was added to medium (Table-XVIII). The production of L-lysine was increased from 55.63 g/l to 62.5 g/l using medium-III, containing 110 g/l glucose plus 15 g/l ammonium sulphate.

TABLE – XVIII: EFFECT OF DIFFERENT CONCENTRATIONS OF BIOTIN ON L-LYSINE PRODUCTION BY CORYNEBACTERIUM GLUTAMICUM FRL No. 3960

CONCENTRATION OF BIOTIN (µg/l)	PRODUCTION OF L-LYSINE (g/l)
25	55.63
50	56.00
75	57.35
100	59.62
125	62.50
150	62.30
175	62.00
200	62.25

Medium used and fermentation conditions:

Medium No. III containing 110g/l glucose and 15g/l ammonium sulphate

pH = 7.2, Temperature = 30°C, Shaking Condition = 200 rpm, Fermentation Time = 96 hours

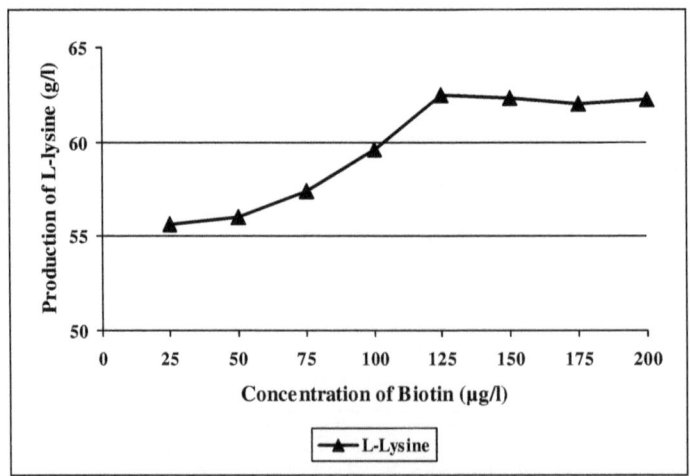

FIGURE – XIV: EFFECT OF DIFFERENT CONCENTRATIONS OF BIOTIN ON L-LYSINE PRODUCTION BY CORYNEBACTERIUM GLUTAMICUM FRL No. 3960

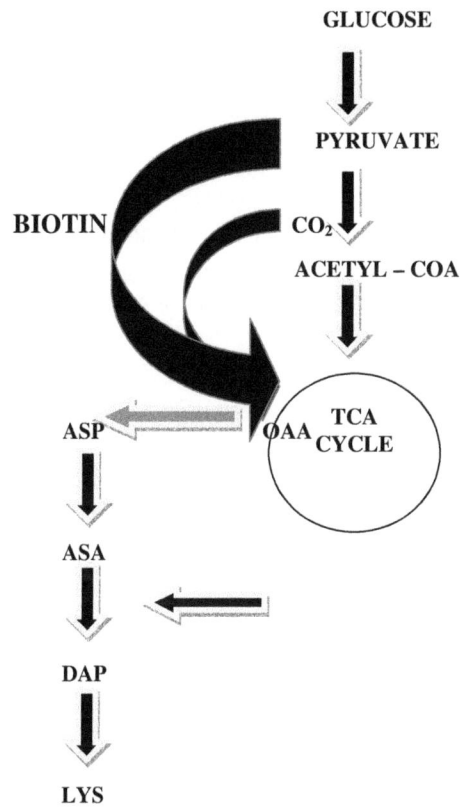

FIGURE – XV: EFFECT OF BIOTIN ON L-LYSINE PRODUCTION

TCA	=	Tricarboxylic acid cycle
OAA	=	Oxaloacetic acid
ASP	=	Aspartic acid
ASA	=	Aspartic β-semialdehyde
DAP	=	Diaminopimelate
LYS	=	Lysine

Effect of vitamin B_1:

Effect of varying concentration of vitamin B_1 was also investigated (Figure-XVI). A significant increase in the L-lysine production was recorded (from 62.50 g/l to 68.65 g/l) when vitamin B_1 was added to medium-III in concentration of 1.0 mg/l (Table-XIX). The medium contained 110 g/l glucose plus 15 g/l ammonium sulphate and 125 µg/l of biotin. T-test was used as statistical method to determine the significance.

TABLE - XIX: EFFECT OF DIFFERENT CONCENTRATIONS OF VITAMIN B_1
ON L-LYSINE PRODUCTION BY CORYNEBACTERIUM GLUTAMICUM FRL No. 3960

CONCENTRATION OF VITAMIN B_1 (mg/l)	PRODUCTION OF L-LYSINE (g/l)
0.1	63.80
0.5	65.50
1.0	68.65
1.5	67.80
2.0	67.50
2.5	68.00

Medium used and fermentation conditions:

Medium No. III containing 110g/l glucose and 15g/l ammonium sulphate and 125 µg/l biotin

pH = 7.2, Temperature = 30°C, Shaking Condition = 200 rpm, Fermentation Time = 96 hours

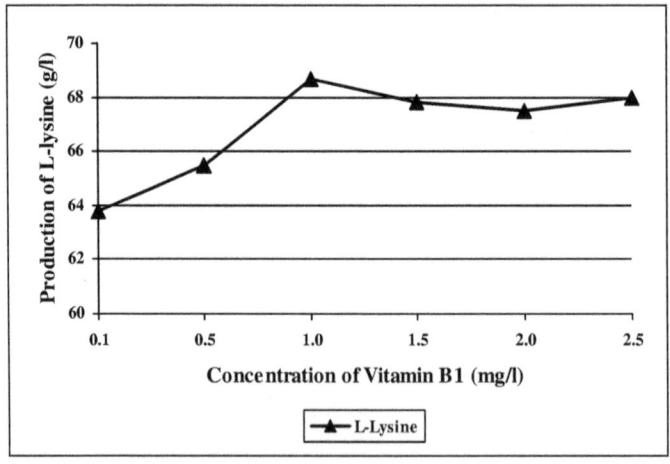

FIGURE – XVI: EFFECT OF DIFFERENT CONCENTRATIONS OF VITAMIN B_1 ON L-LYSINE PRODUCTION BY CORYNEBACTERIUM GLUTAMICUM FRL No. 3960

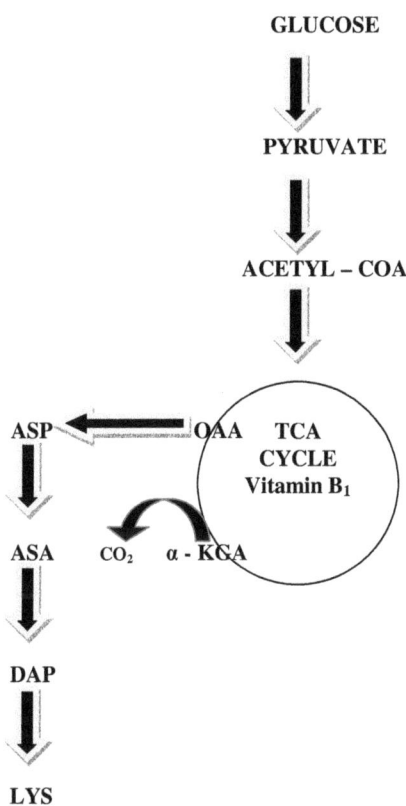

GLUCOSE

PYRUVATE

ACETYL – COA

ASP ◄ OAA TCA
            CYCLE
            Vitamin B$_1$

ASA CO_2 α - KGA

DAP

LYS

FIGURE – XVII: EFFECT OF VITAMIN B$_1$ ON L-LYSINE PRODUCTION

TCA	=	Tricarboxylic acid cycle
OAA	=	Oxaloacetic acid
ASP	=	Aspartic acid
ASA	=	Aspartic β-semialdehyde
DAP	=	Diaminopimelate
LYS	=	Lysine

Effect of aeration :

In order to investigate the effect of aeration on L-lysine production, fermentation was carried out with varying volumes (25 ml to 250 ml) of medium - III. The medium contained 110 g/l glucose, 15 g/l ammonium sulphate, 125 μg/l biotin and 1.0 mg/l vitamin B_1. It was observed that decrease of medium volume in flask (higher aeration) had a significant effect on L-lysine production (Figure - XVIII). An increase in L-lysine production, i.e. 75.65 g/l, 75.20 g/l, 75.00 g/l and 74.75 g/l was recorded when the respective medium volume was kept low (25 ml, 50 ml, 75 ml and 100 ml) in 500 ml flask. Further increase of medium volume (lower aeration) resulted in decreased L-lysine production (Figure - XVIII), despite the longer period of fermentation (Table-XX). A low medium level, i.e. 25 ml, 50 ml and 75 ml in 500 ml flask provided a better areation and thus produced slightly higher quantity of L-lysine than the flask containing 100 ml medium. However, since this could not be adapted as economical method for producing L-lysine on large scale, therefore, a medium volume of 100 ml in 500 ml flask was considered as optimum, producing 74.70g/l L-lysine.

TABLE - XX: EFFECT OF AERATION ON L-LYSINE PRODUCTION BY
CORYNEBACTERIUM GLUTAMICUM FRL No. 3960

MEDIUM VOLUME (ml)	PRODUCTION OF L-LYSINE (g/l)
25	75.65
50	75.20
75	75.00
100	74.75
125	73.68
150	62.50
175	52.65
200	45.00
225	43.55

Medium used and fermentation conditions:

Medium No. III containing 110g/l glucose and 15g/l ammonium sulphate, 125 μg/l biotin and 1.0 mg/l Vitamin B_1

pH = 7.2, Temperature = 30°C, Shaking Condition = 200 rpm, Fermentation Time = 96 hours

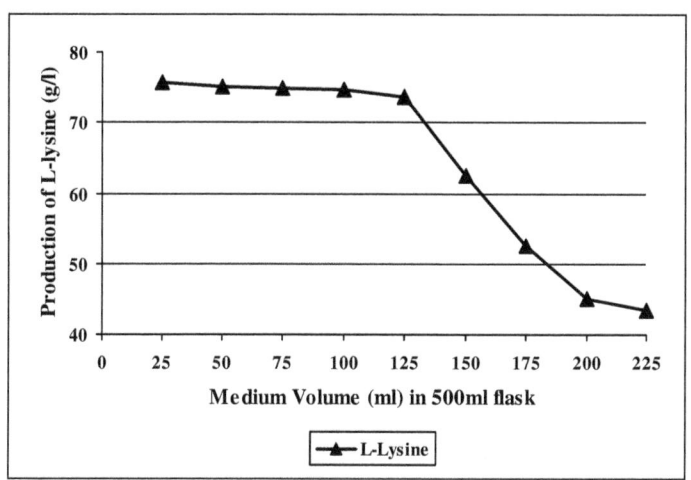

FIGURE – XVIII: EFFECT OF AERATION ON L-LYSINE PRODUCTION BY
CORYNEBACTERIUM GLUTAMICUM FRL No. 3960

4.   LABORATORY SCALE PRODUCTION OF L-LYSINE IN MINI JAR FERMENTERS

Laboratory scale production of L-lysine was carried out using two different types of media, i.e., glucose and molasses media. *Corynebacterium glutamicum* FRL No: 3960 (a homoserine auxotroph and resistant to AEC) was cultivated in minijar fermenters. Cultivation was carried out in two different stages, the shaken-flask inoculum stage and the fermentation stage. The seed culture was cultivated in 250 ml Erlenmeyer flasks, while fermentation was carried out in 2.0 l capacity minijar fermenters, (Eyla - Japan) using glucose (110g/l) and molasses (275 g/l equivalent to 110 g/l as glucose) medium supplemented with inorganic salts for 96 hours. Separation and purification of L-lysine from the fermented broth was carried out by ion exchange method. Strong cation exchange resin, Amberlite IR-120 and weak cation exchange resin, Amberlite IRC-50 were used in the column as adsorbent. After washing and treatment with activated carbon, L lysine was recovered in the form of white crystalline material as L-lysine monohydrochloride. The purity of the material was confirmed by TLC and by infrared absorption spectrum (Figure - XXIII B&C, Section - Discussion). The Infrared (IR) spectral data of the L-lysine optained through glucose and molassess medium is given in section 5.6.2 of Materials and Methods.

90

Both the spectral data were found to the identical with that of the authentic sample of L-lysine. The different peaks in the IR spectra displayed at 2900 cm^{-1}, 1600-1500 cm^{-12}, and 1400-1345 cm^{-1} were assigned due to N-H(stretch broad), C-O and C-N vibrations. The assay of L-lysine was carried out by potentiometric method using perchloric acid.

After separation and purification 67.80 g/l and 65.20 g/l (average of three experiments) crystals of L-lysine monohydrochloride were obtained from glucose and molasses media respectively. The percentage yield ($Y_{p/s}$) of L-lysine for glucose medium was recorded as 0.616 g g^{-1}. (61.6% conversion efficiency) while for molasses medium the percentage yield ($Y_{p/s}$) of L-lysine was 0.592 g g^{-1} (59.2% conversion efficiency). The purity of L-lysine monohydrochloride obtained from glucose and molasses medium was determined to be 99.4% and 98.8% respectively. The process layout of L-lysine fermentation of the present studies is indicated in Figure-XXIV (Section – Discussion).

DISCUSSION

The accumulation of free amino acids in microbial cultures was originally studied from a physiological standpoint under the topic, "extracellular nitrogen compounds (ENC)". Dagley et al. (1950) and Dagley and Johnson (1956) described the excretion of small amounts of amino acids, such as alanine, glutamic acid, aspartic acid and histidine in cultures of *Escherichia coli*. However, the first definite report was that of Kita (1957), who obtained U.S. Patent for L-glutamic acid production using *Cephalosporium sp*. The L-lysine production by fermentation was first reported by Casida and Beldwin in 1956, using a two step method. In the first step, diaminopimelic acid (DAP) was accumulated in the medium containing glycerol and cornsteep liquor by a lysine requiring strain of *Escherichia coli*. Conversion of DAP to lysine was then accomplished with DAP-decarboxylase of *Aerobacter aerogens*. Subsequently, amino acid production by microorganisms continued to attract the attention of various workers throughout the world.

Research on the possible utilization of wild strains revealed that many microorganisms, such as bacteria, yeasts, filamentous fungi and actinomycetes, accumulated amino acids in cultures containing a supplementry source of nitrogen. However, only bacteria showed sufficient productive ability to warrant commerical exploitation. Wild strains from various environments were screened and effective producers of alanine, glutamic acid and valine were isolated, although minor amount of other amino acid were also formed. An excellent classification of the processes for the amino acid production according to the type of microorganism employed was reported by Nakayama, (1972b; Annexure - VII).

The success in industrial production of glutamic acid stimulated further interest in finding strains capable of producing other amino acids as well. However, very soon, it was observed that most wild strains isolated from nature could not produce industrially significant amounts of extracellular amino acids except a few amino acids such as glutamic acid, alanine and valine. The main cause of this fact was the regulation of cellular metabolism, the existence of which has now been well recognised (Nakayama, 1976). For the extracellular production of a desired amino acid, changes in cellular metabolism and/or regulatory controls were required. Thus attempts were made to induce auxotrophic mutants and regulatory mutants. The purpose of utilizing auxotrophic mutants was to negate feedback control mechanisms by limiting the intracellular accumulation of feedback inhibitors or respressors. For example, successful L-lysine fermentation was deviced with homoserine auxotrophic mutants (Kinoshita et al., 1958b; Nakayama et al., 1961a). The purpose of regulatory mutants was to bypass feedback control mechanism by using feedback insensitive mutants, i.e. insensitive to end product inhibition or to end product repression. A variety of such mutants, resistant to amino acid antimetabolites have been induced, for instance, an S-(2-aminoethy)-L-cysteine (AEC) resistant mutants are now in use for L-lysine production because of their capability not to allow conversion of aspartic semialdehyde to L-threonine and thus overproduced aspartic semialdehyde is channelled to L-lysine.

Most amino acids are produced now a days by use of mutants that contain combination of auxotrophic and regulatory mutants. Even more potent amino acid producing strains have been obtained by eliminating the ability of the organism to degrade the product, enzymatic influence and by improving cell permeability in favour of excretion of the product through genetic manipulation / cloning (Bustard and Tryfona, 2005, Ikeda et al., 2006, Hayashi et al., 2006, Anastassiadis, 2007, Wittmann and Becker, 2007, Blombach et al., 2009 and Bulfer et al., 2010). The genes and enzymes involved in L-lysine biosynthesis in Corynebacterium glutamicum are given in Table – XXVII. For these vary reasons, a comprehensive study was carried out to isolate strains of *Corynebacterium glutamicum* from natural habitats in Karachi-Pakistan and an attempt was made to induce combination of auxotrophic and regulatory mutants from *Corynebacterium glutamicum* (L-glutamic acid producing bacteria) for the efficient production of L-lysine.

L-glutamic acid has been recognised as one of the primary product of nitrogen metabolism in the living cell and the enzyme glutamic dehydrogenase system represents important link between the metabolism of amino acids and carbohydrate (Kinoshita et al., 1957a). It was further reported that L-glutamic acid formed by various organisms could easily be transformed into various other amino acids (such as L-lysine, L-methionine, L-threonine, L-isoleucine etc.) and proteinaceous materials. However, this could only be done by inducing auxotrophic and regulatory mutants of the L-glutamic acid producing strains. L-glutamic acid formed in synthetic media containing glucose and inorganic nitrogen source by various microorganisms, have been frequently detected in minute quantities inside the cell or in the surrounding medium (Dagley et al., 1950; Morton and Broadbent, 1955). However, the systemic screening of microorganisms using media containing a sugar (such as glucose) and a nitrogen source (such as urea or an ammonium salt) was first reported in 1957. Two groups of researchers almost simultaneously reported bacteria capable of accumulating large quantities of glutamic acid. First communication came from Asai et al. (1957) of the University of Tokyo, and the other from Kinoshita et al. (1957a and 1957b), who reported for the first time that *Micrococcus glutamicus*, later renamed as *Corynebacterium glutamicum* (Kinoshita et al. 1960) was an excellent producer of L-glutamic acid giving a yield of about 30% and was thus immediately developed for industrial exploitation.

Kinoshita et al. (1957a) reported a comprehensive method for the isolation of L-glutamic acid producing microorganisms (bacteria, streptomyces, yeasts and fungi) from soil, sewage, and animal feaces, etc. About 50 strains of known bacteria and about 600 isolates were tested in the screening experiments. The L-glutamic acid spot on paper chromatogram was detected in the broth of about one fifth strains of known bacteria tested. Many bacteria were found to form a large amount of alanine. The type and amount of amino acids detected varied from one strain to another and each organism revealed its characteristics under the cultural conditions employed. *Micrococcus glutamicus* indicated the highest level of L-glutamic acid accumulation. Other, glutamic acid producer include, *Escherichia coli, Pseudomonas fluorescens, Serratia marcescens, Sarcina lutea*, and *Bacillus megaterium*.

In another study, Kinoshita et al. (1957b) reported screening of 650 strains of bacteria. Approximately 20% of the tested strains showed positive L-glutamic acid production. Once again *Micrococcus glutamicus* gave the highest level of L-glutamic acid accumulation. Other organisms include *Escherichia coli, Bacillus megaterium* and *Sarcina lutea*. *Micrococcus glutamicus* produced 31 g/l of L-glutami found to produce L-glutamic acid: 4.17% of total isolates. The sewage samples analysed in this study were observed more potential than the soil samples. The screening technique adapted in the present study was that described by Daoust (1976) with modification in the composition of screening medium. In the improved screening medium 1.0 μg/l of biotin was added as a growth factor for *Corynebacterium glutamicum* that also permits good L-glutamic acid production. The cultivation and purification of all isolates on a maintenance or storage medium was not carried out as it was certainly not justified in the perliminary stage because their potential to synthesize the amino acid was yet to be determined. Therefore replica-plating technique devised by Lederberg and Lederberg (1952) was used to transfer simultaneously all the colonies from the isolation medium to the screening medium on the same respective site. The detection of L-glutamic acid produced by the replicated cultures was recorded after being exposed to a dose of ultraviolet light, intensive enough to kill the cell and to prevent the growth of L-glutamic acid producing organisms during assay. *Pediococcus*

94

acidilactici ATCC-8042 (previously known as *Leuconostoc mesenteriodes*) was used as test organism. *Pediococcus acidilactici* requires a number of amino acids for its growth and thus could be employed to detect those amino acids.

A suitable dehydrated L-glutamic acid assay medium was used for bio-autography and the growth of the test organism was observed in the immediate area surrounding any colony that produced L-glutamic acid. The intensity of growth (diameter or thickness of the colony) recorded as high (+ + +), medium (+ +), and low (+) which ultimately indicated the formation of L-glutamic acid by each isolate. Results given in Table-I indicates high growth in fifteen (15) isolates, medium in thirty six (36), and low in forty five (45) isolates. The sewage samples were found more potential than the soil samples as seventy four (74) isolates from thirty (30) sewage samples and twenty two (22) isolates from thirty (30) soil samples were found to produce L-glutamic acid indicating ability to warrant commercial exploitation. In total ninety six (96) strains out of twenty three hundred (2300), from sixty (60) samples were found to produce L-glutamic acid.

For quantitative estimation of L-glutamic acid, purification of all ninety six (96) strains was carried out on the same isolation nedium. Thin layer chromatography was used to confirm the production of L-glutamic acid as well as the production of any other amino acid by these strains. The broth of each cultivated strain was applied on aluminium T.L.C. plates, 0.2 mm thick and upflow was made at 25°C. Since the free amino acids have been reported as marked hydrophilic compounds, the separation efficiency was noted with solvent systems, chloroform - methanol - 17% ammonium hydroxide (2:2:1 v/v), n-butanol - acetic acid - water (4:1:1 v/v) and phenol - water (3:1 v/v) (Brenner et al. 1969). However, in the present study, the solvent system n-butanol - acetic acid - water (4:1:1 v/v) provided best separation and was thus used throughout the whole study. Apart from L-glutamic acid, six (6) different amino acids were also identified. These were alanine, aspartic acid, glycine, isoleucine, proline and valine (Table-V). Next to glutamic acid, alanine was found predominent among amino acid. Alanine was produced by fifty one (51) strains, aspartic acid by nine (9) strains, glycine by fifteen (15 strains, isoleucine by thirty two (32) strains, proline by thirty nine (39) strains, and valine by thirty three (33) strains.

Results of the quantitative estimation of L-glutamic acid (Table-VI), was obtained by turbidimetric method (Shockman, 1963) indicated accumulation of L-glutamic acid in significant quantities in ten (10) strains. The highest level of L-glutamic acid production (21.50 g/l) was noted for strain FRL-44, isolated from sewage sample No. 2, followed by FRL-263 (19.8 g/l),

FRL-625 (18.6 g/l), FRL-2025 (18.0 g/l), FRL-962 (17.5 g/l), FRL-86 (16.8 g/l), FRL-170 (15.5 g/l) FRL-1378 (14.4 g/l), FRL-713 (13.5 g/l), and FRL-1125 (13.0 g/l).

Based on the qualitative and quantitative estimations (Table-V and VI respectively) all ninety six strains were screened for *Corynebacterium glutamicum*. A preliminary morphological screening test which was comprised of Gram's staining, acid fast staining and spore staining was first carried out (Table-VII). Strains identified as Gram-positive, non acid-fast, non-spore forming, rod-shaped were then subjected to detailed morphological, cultural, bio-chemical, physiological and chemical studies. Table-VII indicates the results of the perliminary morphological screening. Twentyfive (25) strains out of ninety six (96) were identified as Gram-positive, non acid-fast, non spore forming rod-shaped bacteria. Tables - VIII, IX, X, XI and XII (A, B and C) indicate the results of the detailed morphological, cultural, bio-chemical, physiological and chemical studies respectively. These studies were performed on twenty five (25) strains keeping in view of the preliminary morphological screening test results. For comparative study *Corynebacterium glutamicum* ATCC-13032 was used through out the whole study. Five (5) strains, FRL-44, FRL-54, FRL-263, FRL-625 and FRL-781 were identified as *Corynebacterium glutamicum*. The morphological characteristics of these strains are indicated in Figure - VIII, A, B,C, D and E respectively, while the cultural characteristics of these strains are summarized in Table-XIII. Strains identified as *Corynebacterium glutamicum* were studied according to the system indicated in Bergey's Manual of Systematic Bacteriology (1986). The manual describes *Corynebacterium glutamicum* under the heading of "Irregular, Nonsporing Gram-positive Rods" in Section-15, Vol-2. The descriptions given for *Corynebacterium glutamicum* in this section is of type species ATCC-13032, which were adapted from the studies of the Abe et al. (1967). The Bergey's Manual of Determinative Bacteriology, 9th edition, (1994) also described *Corynebacterium glutamicum* in Group - 19 under the heading of Regular, Nonsporing, Gram-positive Rods.

Kinoshita et al. 1958c presented taxonomical study of glutamic acid accumulating bacteria and proposed a new nomanclature i.e., *Micrococcus glutamicus*. Later, Kinoshita et al. (1960) further studied taxonomical characters of about 20 glutamic acid-producing strains and pointed out 6 basic common characteristic features, i.e., high productivity of glutamic acid in aerobic conditions, Gram-positive, non-sporulating, non-motile, usually ellipsoidal spheres to short rods and biotin requirement. Based on these findings, they proposed that the microorganisms having these 6 common charcteristics may be given a common name or better classified into a new genus. The genus *Corynebacterium* was first named by Lehmann and Neumann in 1896 (Bergey's Manual

96

of Systemic Bacteriology, 1986) for the classification of diphtheria bacillus and a few other animal pathogenic species. However, over many years other nonsporing, irregular, Gram-positive species, both aerobic and anaerobic organisms, were assigned to the genus, until it came to include a very wide collection of ill-assorted organisms (Collins and Cummins, 1986).

The aerobic, Gram-positive, non acid-fast, non-spore forming, rod-shaped bacteria which taxonomically belong to the genera *Corynebacterium, Microbacterium, Cellulomonas, Arthrobacter, Brevibacterium, Curtobacterium*, etc., were also named :Coryneform bacteria" (Jensen ,1952; Yamada and Komagata, 1972b), and "Coryneform glutamic acid-producing bacteria" or more simply "glutamic acid-bacteria" (Kinoshita, 1985) based on glutamic acid producing capability. However, to clarify the differences and similarities among the various strains of glutamic acid producers, several extensive studies have been made. Abe et al. (1967) examined 208 strains of glutamic acid producers. They grouped these strains into 12 types (Table-XXI), 60% of which belong to a single or very closely related species in genus *Corynebacterium*. These organisms were, in general, Gram-positive, non-sporulating, non-motile, ellipsoidal spheres to short rods, pleomorphic, requiring biotin for growth and accumulating large amounts of L-glutamic acid. However, some minor differences were observed in physiological characteristics of these strains. Later Yamada and Kamagata (1972a) reported the morphological, cultural, biochemical and physiological characteristics of 112 strains of Coryneform bacteria. The organisms were divided into seven (7) groups on the basis of combination of the characteristics of Gram-stain, motility, presence of metachromatic granules, pleomorphism, effect of citrate on cell form, acid formation from carbohydrates, assimilation of organic acids, hydrolysis of gelatin, extracellular DNAse, Urease and cell wall composition (Table-XXII). Further a scheme for differentiation of the genera of these bacteria was presented, and problems for classification of Coryneform bacteria were discussed by the same scientists (Yamada and Komagata, 1972b; Komagata et al. 1969; Yamada and Komagata, 1970a and 1970b). They proposed that the Coryneform bacteria could not be classified or identified by the features commonly employed in bacterial taxonomy. Consequently, they adopted some new characteristics such as the mode of cell division, composition of cell wall (principal amino acids and amino sugars) and guanine plus cytosine (G+ C) content in DNA which seemed to be fundamental and basic in character (Table-XXIII).

Various reports published on chemotaxonomy also proved to be important tool for the classification and identification of Corynebacterium (Kinoshita et al. 1960; Keddie and Cure;

1977 and 1978; Keddie and Bousfield; 1980; Suzuki et al. 1981; Uchida and Aida, 1979; Minnikin et al. 1978; Komatsu and Kaneko, 1980; Suzuki and Komagata, 1983 Komura et al. 1975). The genus Corynebacterium has now been well defined on the basis of chemical criteria (Collins and Cummins, 1986). Properties of the members of this genus include a directly cross-linked peptidoglycan based upon meso-diaminopimelic acid (meso-DAP), a arabino-galactan polymer, short chain mycolic acids (25 - 36) carbons, predominantly straight chain saturated and monosaturated fatty acids (occasionally 10 methylbranched acids) and a mol% G + C content of 51 - 60 (Table XXIII).

Table XXIII also provides the primary chemical criteria that can be used to differentiate the genus *Corynebacterium* from other *Actinomycetes* and Coryneform taxa, while Table-XXIV indicates the characteristics differentiating features the species of the genus *Corynebacterium*. On the basis of chemical similarties (cell wall and lipid composition) the genus *Corynebacterium* is most closely related to the genera *Mycobacterium*, *Nocardia* and *Rhodococcus* (Barksdale, 1970), while the genera *Brevibacterium*, *Arthrobacter*, *Microbacterium*, *Curtobacterium, Caseobacter, Cellulomonas, Agromyces,* and plant pathogenic *Corynebacterium* species, are clearly distinguished from other *Corynebacterium,* human and animals pathogens (Table XXIII).

TABLE – XXI: CHARACTERISTICS OF VARIOUS TYPES OF GLUTAMIC ACID PRODUCING STRAINS

PROPERTIES	TYPE	NO OF STRAINS
I. ACID FROM GLUCOSE WITHIN 48 HOURS		
A. UREASE POSITIVE		
1. Nitrites produced from nitrates.		
a. Acid from sucrose.		
b. No acid from salicin.		
c. No acid from mannitol	1	124
cc. Acid from mannitol.	2	3
bb. Acid from salicin		
c. No acid from mannitol	3	8
cc. Acid from mannitol		
d. Acid from maltose.	4	4
dd. No acid from maltose.	5	9
aa. No acid from sucrose.		
b. No acid from salicin.		
d. Acid from maltose.	6	2
dd. No acid from maltose	7	4
bb. Acid from salicin	8	3
2. Nitrites not produced from nitrates	9	6
B. UREASE NEGATIVE		
1. Nitrites produced from nitrates.	10	5
2. Nitrites not produced from nitrates.	11	2
II. ACID PRODUCTION FROM GLUCOSE		
USUALLY DELAYED		
UREASE NEGATIVE	12	38

TABLE – XXII: CHARACTERISTICS TO DIFFERENTIATE SEVEN GROUPS OF CORYNEFORM BACTERIA

Characteristics	Group - 1	Group - 2	Group - 3	Group - 4	Group - 5	Group - 6	Group - 7
Type of cell wall	DL - DAP	DL - DAP	Lysine	Ornithine	Ornithine	L-DAP	DAB
Type of cell division	Snapping	Bending	Bending	Bending	Bending	Bending	Bending
GC content (%)	51-70	60-63	58-65	72-73	66-71	70-72	69-71
Gram stain	Strongly positive	Strongly positive	Weakly positive	Weakly positive Cystite: Strongly positive	Weakly positive	Weakly positive Cystite: Strongly positive	Weakly positive
Motility	Non motile	Non motile	Non motile or motile	Motile or non motile	Motile or non motile	Motile or non motile	Non motile or motile
Metachromatic granula	-	-	-	-	-	-	-
Pleomorphism	Not distinctive	Not distinctive	Remarkable cystite	Fairly distinctive	Weakly	Distinctive Cystite elongation branching	Slightly
Citrate effect	+	±	±	Inhibit	-	-	-
Acid production from sugars	±	-	-	+ Strongly (rapidly)	+ Weakly (slowly)	-	-
Utilization of organic acids	+	+	+ (Widely)	+ (Restricted)	+	+	+
Hydrolysis of gelatin	-	+	+	+ (Slowly)	+ (Slowly)	+	±
DNAse activity	-	+	+	+	+	+	+
Urease activity	+	-	-	-	-	-	-

ADAPTED FROM: Yamada, K. and Komagata, K. (1972b). Taxonomic studies on Coryneform bacteria. V. Classification of Coryneform bacteria. J. Gen. Appl. Microbiol., 18, 417.

GC = Gaunine + Cytosine

DAP = Diaminopemilate

TABLE – XXIII: DIFFERENTIAL CHEMICAL CHARACTERISTICS OF THE GENUS
CORYNEBACTERIUM, ACTINOMYCETES AND CORYNEFORM TAXA

1	2	3	4	5	6	7	8	9
Arthrobacter	L-lysine	A3α	-	59 – 66	-	S, A, I	MK – 9(H2)	(PI)
Brevibacterium	meso - DAP	A1γ	-	60 – 64	-	S, A, I	MK – 8(H2)	(PI)
Cellulomonas	L-Ornithine	A4β	-	71 – 75	-	S, A, I	MK – 9(H4)	(PI)
Corynebacterium	meso - DAP	A1γ	-	51 – 60	22 – 36	S,U (T)	MK – 9(H2) MK – 8(H2)	+
Curtobacterium	D-Ornithine	B2β	-	67 – 75	-	S, A, I	MK - 9	-
Microbacterium	L-lysine	B1α or B1β	+	69 – 75	-	S, A, I	MK – 11, MK - 12	-
Mycobacterium	meso - DAP	A1γ	+	62 – 70	60 – 90	S.U.T	MK – 9(H2)	+
Nocardia	meso - DAP	A1γ	+	64 – 69	46 – 60	S.U.T	MK – 8(H4)	+
Rhodococcus	meso - DAP	A1γ	+	60 – 69	30 – 64	S.U.T	MK – 9(H2)	+
Agromyces	DAB		ND	71 – 76	-	S, A, I	MK - 12	-
Caseobacter	meso - DAP	A1γ	-	65 – 67	30 – 36	S.U.T	MK – 9(H2) MK – 8(H2)	+
Plant pathogenic Corynebacterium	DAB	B2γ	-	67 - 68	-	S, A, I	MK – 9 MK - 10	-

ADAPTED FROM: Bergey's Manual of Systematic Bacteriology (1987), Vol – 2, pp – 1267.

1.Taxon, 2. Major peptidoglycan diamino acid, 3. Peptidoglycan type, 4. N-glycolyl in glycan moiety of walls, 5. Mol % (G+C), 6. Mycolic acid (Carbon atoms), 7. Fatty acid, 8. Major menaquinone isoprenologue (s), 9. Phosphatidylinositol and phosphatidylinosinositol mannoside (s).

Standard symbols: + = 90% or more of strains are positive, - = 90% or more of strains are negative, ND = Not determined, S = Straight chain saturated fatty acids, A = Anteiso-methyl-branched fatty acids, I = Iso-methyl-branched fatty acids, U = Monounsaturated fatty acids, T = 10-methyl-branched acids, ( ) = May be present,

B = Peptidoglycan containing a sery residue in position 1 of the peptide subunit, PI = Phosphatidylinositol,

DAP = Diaminopemilic acid, DAB = Diaminobutyric acid.

Among chemical criteria, the thin layer chromatographic analysis of whole organism methanolysates provides a simple and reliable means of differentiating true *Corynebacterium* (which possess mycolic acids) from the plethora of *Corynebacterium taxa* (viz. *Arthrobacter, Brevibacterium, Curtobacterium, Cellulomonas, Microbacterium*) which lack these characteristic lipids (Goodfellow et al. 1976; Minnikin et al. 1978; Collins et al. 1982a and 1982b; Ariga et al. 1984 and Komagata and Suzuki, 1987). The differentiation of true *Corynebacterium* from the representatives of the genera *Mycobacterium* and *Nocardia* is relatively easy. *Mycobacteria* and *Nocardia* possess relatively large mycolic acids (60 - 90 and 46 - 60 carbon atoms, respectively). *Mycobacteria* and *Nocardia* may also be distinguished from *Corynebacterium* by the fact that the former two taxa possess DNA which is relatively rich in guanine plus cytosine (62-70 mol%) and contain N-glycol residues in the glycan moiety of their walls (Table-XXIII).

Table XII (A, B and C) indicates results of the chemical test performed on strains FRL-44, FRL-54, FRL-263, FRL-625 and FRL-781 identified as *Corynebacterium glutamicum*. All five strains indicated the presence of meso-DAP as major peptido- glycan diamino acid, mycolic acid in the whole organism methanolysates (Table XXV) and G + C (mol%) between 51-60. Furthermore the morphological, cultural, biochemical and physiological characteristics presented in Tables - VIII, IX, X, XI respectively provide comprehensive data to conclude that the strains FRL-44, FRL-54, FRL-263, FRL-625 and FRL-781 be identified as *Corynebacterium glutamicum*. These five strains showed variance with respect to the amount of L-glutamic acid produced by them. The strain FRL-44 produced the highest quantity (21.50 g/l) of L-glutamic acid and thus was selected for the development of mutants by UV-irradiatian and N-methyl-N-nitro-N-nitrosoguanidine (MNNG) treatment.

Natural microbial isolates usually produce commercially important products either in very low concentration or are unable to produce the desired product. Therefore every attempt is being made to produce and to increase the productivity of the desired substance by the chosen organism. Increased yields may be achieved by optimizing the culture medium and growth conditions, but this approach would be limited by the organism's maximum ability to synthesize the desired product. The potential productivity of the organism is controlled by its genome. Hence the genome must be modified to increase the potential yield. The progress in genomics, genetics, biochemistry, physiology, and applications of *Corynebacterium glutamicum* has been reviewed and reported by many researchers (Leuchtenberger, 1996, Hermann, 2003, Ikeda, 2033, Kalinowski et al., 2003 and Eggeling and Bott, 2005). The discovery and cloning of LysE in 1996 (Vrljic et al., 1996) opened the door to genetic manipulation of L-lysine efflux. A comprehensive review on genome-based approach to create a minimally mutated *Corynebacterium glutamicum* strain for efficient L-lysine production was reported by Ikeda et al., 2006. The authors developed a noval approach that employs genomic information to generate an efficient amino acid producer. A comparative genomic analysis of an industrial L-lysine producer with its natural ancestor identified a variety of mutations in genes associated with L-lysine biosynthesis. Among these mutations, the authors identified two mutations in the relevant terminal pathways as key mutation for L-lysine production, and three mutations in central metabolism that resulted in increased titers. These five mutations when assembled in the wild-type genome led to a significant increase in both the rate of production and final L-lysine titer.

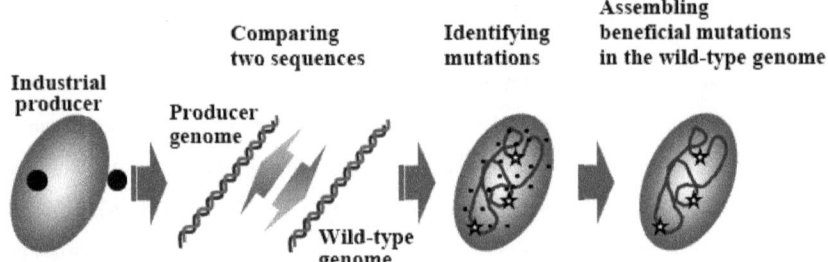

(Methodology to create a minimally mutated strain. Beneficial mutations (stars) relevant to amino acid production are indicated to gether with unnecessary mutations (x).

ADAPTED FROM:

Ikeda, M., Ohnishi, J. and Hayashi, M. . (2006). A genome-based approach to create a minimally mutated *Corynebacterium glutamicum* strain for efficient L-lysine production, *J. Ind. Microbiol. Biotechnol.* **33**. 610 – 615.

In addition, a series of research article on the improvement of L-lysine production using genetic engerining (cloning) have been published by the researchers of Fermentation and Biotechnology Laboratories, Ajinomoto Co., Inc., Kawasaki-ku, Japan (Tsujimoto et al., 2006, Gunji and Yasueda, 2006, Ishikawa et al., 2008a and Ishikawa et al., 2008b).

The cultural requirements of the modified organism was then examined to provide conditions that would fully exploit the increased potential of the culture, while further attempts were also being made to beneficially change the genome of the already improved strain. Thus, the process of strain improvment involves the continual genetic modification of the culture, followed by reappraisals of its cultural requirements. Genetic modification may be achieved by selecting natural variants, induced mutants and recombinants.

The selection of natural variants may result in increased yields but it would be not possible to rely on such improvements because of the heterogeneity of the culture. Therefore, techniques must be employed to increase the chance of improving the culture. It should be kept in consideartion that microorganism usually have regulatory mechanisms which control the amount of metabolites synthesized not exceeding the cell's requirements. To increase the yield of a particular product, it is necessary to induce mutants, in which these regulatory mechanisms are supressed. Such mutants have been classified into two groups; auxotrophic mutants and mutants resistant to analogues (Stanbury and Whitaker, 1984).

A plethora of mutagens have been identified. All produce some kind of unrepairable damage to the DNA leading to mutation. Radiation treatment, chemical modifiers, base analogues and cross linking agents all been shown to have mutagenic effects. However, the efficacy and specificity of the mutagens vary widely (Birge, 1988). Two types of radiations are commonly used for mutation; ultra violet (UV) and X-rays. They differ greatly in terms of the energy involved and hence in their effects. In the present study UV-light was used as a mutagen. The UV-light cause the production of pyrimidine dimers between thymine-thymine, or thymine-cytosine or cytosine-cytosine pairs. The UV-irradiation may also result in distortion of the backbone of the DNA helix causing replication errors. Evidence has been accumulated suggesting that UV mutagenesis in both eucaryotes and procaryotes is an enzymatic process, arising as a result of errors made during the repair of damaged DNA (Moat, 1979).

It is believed that primary effect of the chemical modifiers is due to their ability to deaminate cytosine and guanine. The deamination results in changes in the hydrogen bonding relations so that, at the next replication, adenine or thymine, instead of guanine or cytosine, is inserted. Some moderately specific mutagens, such as hydroxylamine reacts primarily with cytosine. It may also attack uracil or adenine. In the present study an alkylating agent N-methyl-N-nitro-N-nitrosoguanidine (MNNG) was also used as a mutagen. The alkylating agents produce mutagenic activity by attaching ethyl or methyl groups at the 6-position of the purine ring, resulting in the mispairing of the base. The MNNG, having extremely potent mutagenic effect, produces 6-methylguanine thus exhibiting greatest effect at the replication fork (Birge, 1988).

A L-lysine structural analogue S-(2-aminoethyl)-L-cysteine (AEC) was employed in the present study to obtain analogue resistant mutants. AEC, a sulphur analogue of L-lysine was reported to inhibit the growth of *Lactobacillus mesenteroides, Lactobacillus arabinosus, Escherichia coli, Bacillus subtilis, Brevibacterium flavum* (Sano and Shiio, 1970) and *Corynebacterium glutamicum* (Nakayama and Araki, 1973). AEC also reported to inhibit the activity of an enzyme aspartate kinase (Sano and Shiio, 1970). Therefore derivation of mutants resitant to AEC was attempted with *Corynebacterium glutamicum* FRL-2753 (a homoserine auxotroph) and their L-lysine production was examined. Since the use of AEC alone produced weak inhibitory effects, L-threonine (one of the strongest stimulator) was added in the medium. With the addition of L-threonine in the medium, the inhibitory effect was markedly enhanced.

Isolation of auxotrophic mutants in the present study was carried out by penicillin selection method. The choice of the method is based on the fact that penicillin was more effective in killing cells that were actively dividing (Moat, 1979). A large number of nonmutants (wild type) cells were eliminated by this method, increasing the possibility of finding a few mutants in a large population. Although the chance of finding mutants may be increased substantially. It should be emphasized that obtaining a suitable number of auxotrophic mutant may still require testing of considerable numbers of isolates. About 30 to 40 colonies were tested on one plate. With a view to avoid contamination, such that the doubtful colonies were confirmed by microscopic observation. The screening of required growth factor(s) of the mutants was carried out by auxanographic technique using filter paper disc (3 mm in diameter) soaked in different amino acids and vitamins.

Mutation of cultures undoubtedly has played a major role in the production of amino acids. To improve the yield of an amino acid, mutants having multiple markers(including auxotrophs and analogue resistant mutants contributing to the production of the designated amino acid) were highly desirable. The multiple markers also contribute to a higher yield by stabilizing the productivity against back mutation during fermentation. Now a days from auxotrophic mutants serval amino acids are being produced, including L-lysine, L-citruline, L-leucine, L-ornithine, L-proline, L-threonine, L-tyrosine etc.The active attempts to utilize the phenomenon of auxotrophy for the industrial production of amino acids was started during late 1950s. The first prominent result was obtained in producing diaminopimelic acid using a lysine requiring auxotroph of *Escherichia coli* (Casida and Baldwin, 1956). The diaminopimelic acid was then decarboxylated to produce L-lysine in the second step (Figure - XIX).

Most of the high L-lysine yielders were auxotrophs for one or more amino acids. The amino acids required by such auxotrophs were not always related to the L-lysine pathway. Infact they have no apparent biochemical relationship with lysine pathway. Since the discovery of a relatively high L-lysine producing homoserine dependent mutant of *Corynebacterium glutamicum* (Kinoshita et al., 1958b), many other bacteria with similar mutational block (Figure - XX) have been used commercially for L-lysine production. Both auxotrophic mutants requiring amino acids and regulatory ones selected as mutants resistant to amino acid analogues have been proved to be suitable L-lysine producers (Tosaka and Takinami, 1986). Apart from *Corynebacterium glutamicum*, high L-lysine production was also reported in auxotrophic mutants of *Escherichia coli* and *Bacillus subtillus* (Mahmood et al., 1996).

The biosynthetic block most conducive to effect L-lysine accumulation was that requiring homoserine or threonine plus methionine for growth. Specially the homoserine auxotrophs of *Corynebacterium glutamicum* have been reported as heavy L-lysine excretors (Kinoshita et al., 1958b). Other mutants such as the threonine, methionine, and isoleucine auxotrophs were also reported as producers of L-lysine, but inferior to the homoserine auxotrophs (Nakayama et al., 1961a). However these reports did not deal with sufficient number of mutants so as to compare the producing ability of L-lysine satisfactorily. L-lysine production by homoserine, threonine, methionine and isoleucine auxotrophs of *Brevebacterium flavum* was also reported by Sano and Shiio (1967). They reported isolation of more than 100 mutants relating to L-lysine biosynthesis from *Brevebacterium flavum*. They further reported that, though the homoserine auxotrophs of *Corynebacterium glutamicum* were the most powerful producers of L-lysine, as however, some auxotrophs gave fairly different data. The methionine and isoleucine auxotrophs produce little amount of L-lysine, while the threonine auxotrophs were separated into two groups as powerful and weak producers of L-lysine. In the present study, in order to produce and to improve the yield of L-lysine by one of the isolate *Corynebacterium glutamicum* FRL No. 44, two genetic approaches were undertaken. The first approach was through the isolation of an homoserine auxotrophic mutant, while the second approach via the isolation of mutants resitant to the L-lysine analogues AEC. Therefore in order to induce a feedback resistant mutant, AEC, a toxic analogue of L-lysine (Figure-XXII) was employed. The AEC, in cooperation with L-threonine has been reported to exert a false feedback inhibition on aspartate kinase,and therefore the parent strain could not grow in their presence. In addition, the AEC resistant mutants have desensitized aspartate kinase which is quite free from feedback inhibition by L-lysine plus L-threonine. This property of AEC resistant mutants favour the overproduction of L-lysine (Hirose and Shibai, 1980). Schrumpf (1991) and Schrumpf et al. (1992) also reported aspartate kinase as the rate-limiting step in the production of L-lysine by *Corynebacterium glutamicum*.

Tosaka et al. (1978 c) classified the L-lysine producing microorganisms into three groups. The first type was a homoserine or threonine and methionine requiring mutant. The second type was threonine or methionine sensitive mutant and the third type was a mutant resistant to L-lysine analogues, such as S-(2-aminoethyl)-L-cysteine (AEC). The regulatory control mechanism of L-lysine biosynthesis in *Corynebacterium glutamicum* is very simple. (Figure-XXI), the enzyme aspartate kinase (E.C 2.7.2.4) which converts aspartate to ß-aspartyl phosphate has been reported to be inhibited by lysine plus threonine (Shiio and Sano, 1969 and Hirose and Shibai, 1980). Thus the homoserine auxotrophic mutants resistant to AEC overproduces L-lysine only if the concentration of threonine was kept below a certain optimum level (Figure-XI). Auxotrophy and resistance to feedback inhibition, when genetically combined into a single strain, resulted in increased production of L-lysine. Sano and Shiio (1970) first reported the isolation of mutants of *Brevibacterium flavum* whose aspartate kinase was resistant to feedback inhibition by lysine plus threonine. This feedback inhibition was achieved by selecting mutants resistant to S-(2-aminoethyl)-L-cysteine (AEC). Therefore, increased production of L-lysine by the above mentioned mutants might be due to the starvation of threonine. This results in the decrease of the feedback inhibition of aspartate kinase which was the first key anzyme of L-lysine biosynthesis. If aspartate kinase was itself genetically altered to become resistant to the concerted feedback inhibition by threonine plus lysine, L-lysine biosynthesis would not be affected by threonine and lysine, and, therefore, much L-lysine would be produced independently of the concentration of threonine. Sur et al. (1991) reported the use of mutants deficient of amino acids for the production of L-lysine by fermentation. They reported the treatment of a homoserine auxotrophic mutant of *Corynebacterium glutamicum* with nitroguanidine, and the isolation of auxotrophic mutants requiring isoleucine. The double auxotrophic mutant was genetically stable and produced more L-lysine (56 mg/ml in 8 days) than the parent strain.

The work reported in this book, includes the isolation of homoserine auxotrophic mutants of *Corynebacterium glutamicum* from a wild strain (FRL No. 44) and development of mutants whose aspartate kinase was genetically altered and made resistant to feedback inhibition (Figure - XXI). The comparative studies on L-lysine producing abilities of auxotrophic mutants derived from *Corynebacterium glutamicum* FRL No. 44 (Parent strain), were also carried out. Among homoserine, methionine plus threonine, methionine, threonine, leucine and isoleucine auxotrophs, the homoserine auxotrophic mutant FRL No. 2753 was the best producer of L-lysine (producing 26.4 g/l L-lysine) accumulating none of other amino acids in glucose medium supplemented with inorganic salts. The strain FRL No. 2753 along with other ten (10) homoserine auxotrophic mutants required vitamin B_1 as growth factor. The strain was selected for further improvement by developing a series of mutants resistant to S-(2-aminothyl)-L-cysteine (AEC).

The AEC mutants of *Corneybactrium glutamicum* FRL N. 2753 showed great variance in their ability to produce L-lysine. Only a few mutants were noted superior in their ability to produce L-lysine. Five (5) mutants, out of eighty five (85), were found to produce L-lysine significantly higher than the parent strain (FRL No. 2753). Eight (8) mutants produced L-lysine slightly higher than the parent strain, six (6) mutants produced less and in seven (7) mutants the ability to produce L-lysine remained undchaned, while fifty nine (59) mutants produced slightly higher quantities of L-lysine than the parent strain FRL No. 2753. The highest production of L-lysine (33.0 g/l) was recorded in strain FRL No. 3960 and was selected for further studies. The aspartate kinase of this strain was resistant to feedback inhibition. Furthermore, the results presented in Table-XVI and Figure-XXI, confirmed that the site of mutation in strain FRL NO. 3960 was not the permeability of the cell wall but the enzyme, aspartate kinase. The fact that the strain FRL NO. 3960 continued to produce L-lysine for 120 hours (Table-XVII) was indicative of the stability of the strain.

The genetic induction of an auxotrophic or a regulatory mutant that can overproduce amino acid, is of course extremely important. However at the same time the process optimization (especially, the media composition, addition of certain growth factors and culture conditions) is also indispensable for the scale-up and commercial production of amino acids for increased productivity, elimination of undesirable co-metabolites, better consumption of nitrogen and carbon sources and separation final product from the fermented broth.

In the present study a precise modification in media composition was carried out by increasing the glucose (carbon source) and ammonium sulphate (nitrogen source). The concentration and the type of sugar used is the most significant factor in determining the production and yield of L-lysine as well as other amino acids. The biosynthesis of L-lysine by *Corynebacterium glutamicum* is reported to be a branched pathway that also produces methionine, threonine and isoleucine

(Figure-XVII). A number of intermediates, such as pyruvate, oxaloacetate and aspartate are also synthesized in various ways depending on the substrate which acts as carbon source (Hassan and Rogers, 1990). The most satisfactory way of determining the optimum level of carbon source, was reported by Daoust, 1976, for the production of amino acids. The method includes a numbers of shake flask experiments with medium containing different concentrations of sugar supplemented with sufficient quantity of nitrogen source to satisfy the demand imposed by increasing the carbon levels. An analysis of the amount of amino acid produced and the amount of sugar remaining after the process provides data necessary to calculate the percent yield of the product. The sugar concentration which provides the maximum percent yield of the product should be in the optimum concentration so as to exhibit maximum efficiency.

A comprehensive description of factors in order to improve the fermentation process and L-lysine yield was reported by Anastassiadis, 2007. These include, increase in the microorganism's intrinsic productivity characteristics by classicial mutagenesis and genetic engineering, optimization of media composition, rate of aeration and agitation, temperature, pH and CO_2 control and downstream processing. The author (Anastassiadis, 2007) considered, medium composition as very important factor strongly influencing fermentation processes, often being object of extensive process development and optimization studies. The culture medium must satisfy in a suitable manner the requirements of microbial growth and production. Defined media acquiring pure growth requiring nutrients and essential additives or alternatively undefined media containing natural organic substance such as soybean hydrolyzate, cornsteep liquor, yeast extract or peptone are used for L-lysine fermentation (Anastassiadis, 2007).

Use of starch as an efficient carbon source is also reported (Tateno, et al., 2007). The authors achieved direct conversion of starch to L-lysine from *Corynebacterium glutamicum* by displaying α- amylase on its cell surface. Using such a cell surface display system, the enzyme used could be easily be recycled, hence, this noval system usng Corynebacterium glutamicumm can be applied in a wide variety of processes.

The choice of the nitrogen source depends essentially upon the nutritional characteristics of the production culture. It has been reported that the auxotrophic mutants require small amounts of organic nitrogen (such as urea and protein hydrolysates) as compared to inorganic nitrogen (such as ammonium sulphate and ammonium nitrate). However, specific requirements of the mutant cultures must be kept in consideration (Daoust, 1976). Ammonium sulphate and urea have been reported to be the best inorganic and organic sources of nitrogen respectively for L-lysine production (Sassi et al., 1990).

In the present study maximum yield of L-lysine was recorded with increasing concentration of glucose and ammonium sulphate upto 110 g/l glucose and 15 g/l of ammonium sulphate added to medium-III (Table-XVI). At this concentration negligible quantity of residual glucose was noted after 96 hours and 120 hours of fermentation. The L-lysine production at this concentration was recorded as 55.63 g/l with corresponding yield (Y p/s), i.e., 0.5057 g L-lysine per unit weight of glucose (g g^{-1}) after 96 hours. It was also observed that the yield (Y p/s) of L-lysine started decreasing after 110 g/l of glucose and 15 g/l of ammonium sulphate. This may be ascribed to substrate inhibition and limitation of certains nutrients (Hassan and Rogers, 1990) or the diversion of substrate to some undesirable by products (Misra et al., 1980). These factors are also responsible for the retention of significant amounts of unused (residual) glucose in the culture broth, especially in the flasks containing 130 g/l, 150 g/l and 170 g/l of glucose. Hence, the yield of L-lysine was well below the optimum level. As expected, use of more glucose would produce more L-lysine but not necessarily increasing the yield of L-lysine (Y p/s). Another finding is that, the increase in the amount of ammonium sulphate from 15 g/l to 30 g/l does not significantly increase the production of L-lysine. These findings were inconsistence with the results of Hassan and Rogers (1990) indicating that an increase of ammonium sulphate in the medium (after optimum level) does not affect the L-lysine production significantly.

During the course of fermentation, it was also observed that with the increase in glucose concentration from 70 g/l to 110 g/l and fermentation time from 36 hours to 96 hours, a significant increase in biomass (dry cell weight) was recorded. It is apparent from Table-XVII that the L-lysine production is growth associated. This was further confirmed by plotting L-lysine production vs biomass (Figure-XIII). This is in contrast to the findings of Zaitseva et al. (1973) who reported L-lysine fermentation as a biphasic process. The first phase was characterized by an intense accumulation of biomass. The culture in this phase consisted primarily of large basophilic cells. In the second phase, the growth rate of the cells decreased and the lysine concentration in the medium increased. The cells revealed a low basophilia.

When glucose concentration was kept 110 g/l with a fermentation time of 96 hours, 14.45 g/l biomass was recorded from the medium-III with negligible residual glucose. Production of L-lysine at this concentration was recorded as 55.63 g/l with a yield of 0.5057 g L-Lysine per unit weight of glucose (gg^{-1}). At this stage it was noted that 1g of biomass produced 3.849 g L-lysine. In a similar study Misra et al. (1980) reported 3.36 g L-lysine from 1g biomass of *Corynebacterium glutamicum*, while Sassi et al. (1990) reported 2.96 g L-lysine/g of biomass using *Corynebacterium glutamicum*. These two values are lower than the value exhibited by strain FRL No. 3960. However, with further increase in glucose concentration from 110 g/l to 170 g/l, no significant change in biomass was observed even after 120 hours of fermentation. Infact a significant amount of residual glucose was noted with the increase in glucose concentration from 110 g/l to 170 g/l (Table-XVI), inspite of the addition of sodium acetate in the media which is reported to increase the conversion ratio of consumed sugar to L-lysine (Wang et al., 1991). From these results it can be concluded that the optimum concentration of glucose and ammonium sulphate were 110 g/l and 115 g/l respectively.

The role of biotin, in amino acid fermentation was first observed during the course of detailed studies on L-glutamic acid fermentation. It was reported that biotin has a characteristic role in the oxidation of glucose in the synthesis of proteins as well as in cell permeability (Tosaka et al., 1979b). The enzymatic role of biotin as a CO_2 carrier of covalent bond is now unequivocally established (Moss and Lane, 1971). A biotin containing enzyme might function to provide oxaloacetate and consequently aspartate for both the operation of the Tricarboxylic acid cycle (TCA-cycle) and the biosynthesis of cell constituents (Tosaka et al., 1979 b; Figure-XV). Aspartate has been established as an intermediate leading, to the synthesis of L-lysine (Tosaka and Takinami, 1978). Tosaka et al. (1979 b) also reported that glucose is converted to pyruvate which enters into the TCA-cycle and forms oxaloacetate and eventually L-lysine. Carbon dioxide fixation into pyruvate is an important route to supply C_4-dicarboxylic acids which are necessary for the biosynthesis of glutamate and aspartate. On the basis of results obtained, they suggested that carbon dioxide fixation might occur in the presence of excess biotin (Figure-XV). In L-lysine fermentation, biotin is usually added into culture medium at a level of 50 µg/l of medium. Tosaka et al. (1979 b) reported the correlation between L-lysine production and level of biotin in culture medium in the range of 50 µg/l to 500 µg/l. This effect was observed only when glucose or pyruvate was used as substrate. Further more the addition of excess biotin led to a significant reduction in the amount of other amino acids such as L-alanine, L-valine and L-Leucine, which are amino acids of pyruvate family. This indciates that pyruvate could be converted to L-lysine preferentially in the presence of excess biotin. In the present study, the optimum level of biotin was recorded as 125 µg/l. As shown in Table-XVIII, production of L-lysine was significantly increased from 55.63 g/l to 62.50 g/l (12.35%) in presence of 110 g/l glucose and 15 g/l ammonium sulphate. A further increase in biotin concentration had no effect on the production of L-lysine (Figure-XIV). The increase in L-lysine production was inconsistent with numerous previous reports and may be ascribed to the stimulation of pyruvate carboxylase by biotin which consequently lead to increase of aspartate formation through the increased oxalo-acetate formation.

Like biotin, vitamin B_1 in the form of thiamine diphosphate also plays significant role in the oxidation of glucose. The role of thaimine diphosphate in the oxidative decarboxylation of pyruvate and a-ketoglutarate in bacterial system is well established (Koike and Reed, 1960). The oxidative decarboxylation of pyruvate leads to increased formation of acetyl CoA while that of a-ketoglutarate to succinyl CoA and CO_2 and ultimately the oxaloacetate. In the present study a significant increase in the production of L-lysine from 62.50 g/l to 68.65 g/l (8.96%) was recorded when vitamin B_1 in a concentration of 1.0 mg/l was added to medium-III containing 110 g/l glucose and 125 µg/l biotin (Table-XIX)., A further increase in vitamin B_1 cencentration had no significant effect on the production of L-lysine (Figure-XVI). The increase in L-lysine production may be ascribed to the effect of vitamin B_1 which acts as a Co-enzyme in the TCA-cycle for the oxidative decarboxylation of pyruvate as well as a-ketoglutarate, leading to increased production of oxaloacetate and consequently the aspartate (Figure-XVII). The aspartate has been well established as an intermediate leading to the synthesis of L-lysine (Tosaka and Takinami, 1978).

It has been reported that sufficient supply of oxygen to satisfy the cell's oxygen demand is essential for the maximum production of L-lysine (Akashi et al., 1979). In the present study to increase the concentration of oxygen, fermentation was carried out with varying volumes (25 to 250 ml) of medium-III (containing 110 g/l glucose, 125 µg/l biotin and 1.0 mg/l vitamin B_1), in 500 ml Erlenmeyer flask. It was observed that a decrease in medium volume i.e, increased aeration had significant effect on the production of L-lysine (Table-XX). A medium volume of 100 ml in 500 ml flask shared optimum level, producing 74.70 g/l L-lysine. A medium volume below 100 ml in 500 ml flask contributed vary slight increase in the production of L-lysine. However, an increase in the medium volume above 125 ml (lower aeration) resulted in significantly decreased production of L-lysine (Figure-XVIII) despite the longer period of fermentation. These results are in consistence with the pervious reports of Nhan et al. (1976) and Hanel et al. (1981). An increase in L-lysine production due to increase oxygen may be ascribed to the high activity of TCA-cycle

(Hassan and Rogers, 1990). The large volume flasks with low medium level no doubt provide better aeration due to bigger surface area to volume ratio. However this could not be adopted as an economical method for producing L-lysine on large scale. Thus a medium volume of 100 ml in 500 ml flask was considered as optimum, producing 74.70 g/l L-lysine.

Recently, L- lysine production by *Corynebacterium glutamicum* was improved by metabolic engineering of the TCA cycle. The 70 % decreased activity of isocitrate dehydrogenase, achieved by start codon exchange, resulted in a more than 40 % improved lysine production. By flux analysis this could be correlated to a flux shift from TCA cycle towards anaplerotic carboxylation (Becker et al., 2009).

Industrially, L-lysine is mostly produced by fermentation technique. Coello et al. (1992) reported L-lysine as the third highest amino acid production on a large industrial scale. Commercially, the largest amount 80% is produced by fermentation and 20% still by chemical synthesis. The related merits of either process will to a large extent depend upon the in depth expertise and efficiency developed within individual companies. The major raw materials which are generally employed are very typical of the requirements for other fermentation industries, namely starch, sugar, molasses, plant hydrolysate etc. These raw materials are of agricultural origin and are abundantly available in Pakistan. Thus as such, the lack of capability to produce L-lysine at industrial level in Pakistan is not due to the nonavailability of raw materials, but the absence of appropriate fermentation technology, i.e., the lack of suitable strains of bacteria and the lack of development of suitable fermentation process.

The L-lysine fermentation is a complex microbiological process, influenced by several biochemical and physical parameters. Therefore, understanding of the physiology and biochemistry of amino acid producing strains are highly essential in order to choose the most suitable physio-chemical conditions for their maximum performance in productivity. Commercially, L-lysine is usually produced by batch fermentation, and most of the studies of the physiology of L-lysine production have also been carried out by batch procedure (Nakayama, 1985). Studies on L-lysine production in continuous culture have also been made, the aim was to improve the production process and to compare it with the productivity of the batch process.

L-lysine is produced exclusively by mutants of *Corynebacterium glutamicum* obtained by classical breeding. While good production strains have been obtained, the characters required for high productivity still remain largely unknown. Much effort has recently been devoted to elucidate the mechanisms of microbial production of L-lysine. The most outstanding results concern metabolic regulation and transport of L-lysine. The biosynthetic pathways of L-lysine are now well known, and the focus of attention has therefore been moved to metabolic control and its breakdown. In the normal metabolism of microorganisms, L-lysine synthesis is in equilibrium with requirements, i.e., it is carried out under limiting conditions. The accumulation of large quantity of L-lysine is therefore a pathological phenomenon arising from an artificial distortion of metabolism in the strain. In this respect, L-lysine production is essentially different from traditional catabolic fermentations.

Considerable emphasis has been given to the theoretical and practical problems of L-lysine fermentation including selection of carbon and nitrogen source and importance of different factors, such as addition of nutrients (organic and inorganic) and rate of aeration and agitation. The major problems regarding raw materials for L-lysine fermentation involve carbon source which produce the structural frames, and energy sources for the fermentable microorganisms. For industrial production of L-lysine, the choice of raw materials depends largely on economic considerations. This does not means simply the cost of raw materials, but also the cost for isolation and purification of fermented broth as well as various considerations concerning fermentation yield, fermentation hours, treatment of waste materials produced by fermentation and so on. Similarly the optimization of culture conditions in relation to oxygen supply and carbon dioxide removal has played a key role in the scale-up and commercial production of L-lysine. These factors determined both the rate of cells growth and product formation. An interesting article relating to manufacture of stabilized brown juice for L-lysine production on large scale has also been reported (Thomsen et al., 2004). Using this concept, it is possible to supply the L-lysinr manufacturers with stabilized high quality brown juice all year round for industrial production of feed grade liquid L-lysine concentrate.

Extraction and purification of L-lysine from the fermented broth is also a key step in L-lysine fermentation. Ion exchange resins have been widely applied for the extraction and purification of L-lysine from fermented broth (Samejima, 1972 and Nakayama, 1985). However electrodialysis and solvent extraction were also reported but mostly applied for extraction and purification of L-glutamic acid (Samejima, 1972). The adsorption and elution characteristics of L-lysine on ion exchange resins are dependant upon various physio-chemical factors such as pH, the ionic dissociation of amino acid and impurity of ions, ionic strength, etc. By making use of various combinations of such factors, ion exchange resins can be utilized for concentration, mutual separation, desalting and so on. In the present study a special acid resistant column was used for this purpose. An additional facility to control the rate of flow of supernatant was provided in the column. The adsorption of L-lysine by ion exchange resine was greatly affected by the pH of the solution. It was observed that removal of L-lysine from fermented broth was most efficient when the fluid was adjusted to pH 2.0 before passing through the ion exchange resin column. At this pH, competition by Ca^{2+}, K^+, Na^+, and NH_4^+ for the resin was reported minimum (Muradyan et al., 1980).

For the fermenter scale production, two different types of media were formulated. Medium I, containing 110 g/l glucose, while medium II was prepared from 275 g/l of cane molasses, containing 40% reducing sugar. Both media contain biotin, however medium I contains a higher concentration, 125.0 μg/l than medium II. Since, molasses contains enough biotin, therefore, the concentration of biotin in medium II was kept low (25.0 μg/l). Keeping in view the high cost of organic nutrients, such as yeast extract, peptone, and meat extract, they were replaced by low cost and locally available raw materials such as corn steep liquor, soy been protein acid hydrolysate and fish meal respectively in medium II. Concentration of inorganic salts were kept similar in both media to those used in shake flask experiments. Cane molasses are generally used as carbon source in the industrial production of L-lysine (Nakayama, 1972a; Misra et al., 1980; Chancharoensin and Bhumiratana, 1983; Nakayama, 1985; Sassi at al., 1990 and Wibowo et al., 1992). However cane molasses contains many constituents which are toxic to bacterial growth. Therefore superphosphate treatment was done to remove the toxic substances in the form of heavy precipitate.

The use of cornsteep liquor, soyabean protein acid hydrolysate and fish meal have been reported as organic nutrient source for L-lysine production by many workers (Nakayama et al., 1979; Inuzuka and Hamada, 1976; Kurihara et al., 1972 and Nakayama and Araki, 1973). These organic nutrients not only accelerates the rate of growth of microorganisms but also increase the production of L-lysine (Sassi at al., 1990). In addition they are also reported as useful source for threonine and methionine (Nakayama, 1972a). From the results of the laboratory scale study, conclusion can be drawn that locally available cane molasses (after superphosphate treatment) is a useful carbon source and soyabean protein acid hydrolysate, cornsteep liquor and fish meal are useful organic nutrients for L-lysine production. A simple analysis of the cost of the raw materials used for the production of L-lysine is presented in Table - XXVI. Based on the average amount of

L-lysine produced by both media (67.80 g/l vs 65.20 g/l), the cost of raw material was US$ 1.05 vs US$ 0.27 per kg. The purity of L-lysine recorded for glucose and molasses medium were 99.4% and 98.8% respectively. The quantities of L-lysine produced by both media as well as the purity of the products are quite comparable. Thus keeping in view the high cost of glucose, the cane molasses justified its use as a carbon source for L-lysine fermentation. Similarly, the use of soyabean protein acid hydrolysate, cornsteep liquor and fish meal are also justified because of the high cost of yeast extract, peptone and meat extract respectively. Hence, a considerably less expensive L-lysine fermentation could be carried out using these raw materials. The most interesting feature of the present study was the percentage yield of L-lysine. Some of the previous reports on *Corynebacterium glutamicum*, using molasses and glucose medium reflect highly variable results. Nakayama and Araki (1973) reported the production of 39.5 g/l of L-lysine using cane molasses medium containing 10% reducing sugar. Later, Nakayama and Araki (1981) reported 42 g/l L-lysine on cane molasses medium containing 10% reducing sugar. Nakayama (1985) further reported 44 g/l of L-lysine on cane molasses (containing 20% glucose) after 60 hours with a yield of 40%. Smekal et al. (1988) obtained 33 to 42 g/l of L-lysine with a conversion rate of 26% in 96 hours using molasses and starch hydrolysates. A high yield of L-lysine, 52 g/l on cane molasses after 96 hours was obtained by Yonekura et al. (1988). Similar results were published by Hilliger et al. (1990). They reported 52.4 g/l of L-lysine from 290 g/l of beet molasses after 70 hours.

The pattern of L-lysine yield, on glucose medium, is also more or less similar to that on molasses medium. Misra et al. (1979) reported 45 g/l of L-lysine on glucose medium after 120 hours with a yield of 45%. Schrumpf et al. (1992) reported 44 g/l of L-lysine from 100 g/l of glucose. Some of the less elaborated results indicated a very high L-lysine productivity. Plachy and Ulbert, (1988) reported 60 g/l of L-lysine in 60 hours with a conversion rate of 28%. Similarly, Hilliger et al. (1991) reported production of 62.2 g/l of L-lysine. Recently Won et al. (1990) reported a very high yield of L-lysine i.e., 120 g/l. The production of L-lysine by immobilized cells and by continuous culture was also reported to be very high. Nasri et al. (1989) reported 60 g/l of L-lysine in 120 hours with immobilized cells using glucose medium, while Hirao et al. (1989) obtained 105 g/l of L-lysine through continous culture using glucose. The strain, *Corynebacterium glutamicum* FRL No. 3960, (a homoserine auxotroph-resistant to AEC) which was used in the present study, produced significant quantity of L-lysine, i.e., 67.80 g/l and 65.20 g/l on glucose and molasses medium respectively. The yield ($Y_{p/s}$) for glucose medium was recorded as 0.616 gg^{-1} (61.6%

conversion efficiency), while for molasses medium it was 0.592 gg^{-1} (59.2% conversion efficiency) with a purity of 99.4% and 98.8% respectively.

TABLE - XXIV : CHARACTERISTICS DIFFERENTIATING THE SPECIES OF THE GENUS CORYNEBACTERIUM

Characteristics	1	2	3	4	5	6	7	8	9	10	11	12	13	14	15	16	17	18
ACID PRODUCED FROM																		
Glucose	+	+	+	-	+	+	+	+	+	+	+	+	+	+	+	+	+	-
Arabinose	+	d	-	-	-	ND	-	-	-	-	-	-	-	ND	-	-	d	-
Xylose	-	-	-	-	-	-	-	-	-	-	-	-	-	ND	-	-	-	-
Rhamnose	+	+	-	-	-	ND	-	-	-	-	ND	-	-	ND	-	-	-	-
Fructose	+	+	+	-	+	+	+	+	+	+	ND	+	+	+	+	+	+	+
Galactose	+	+	+	-	-	ND	d	+	-	-	-	-	+	+	-	-	+	-
Mannose	+	+	+	-	+	d	d	+	-	-	-	+	+	+	+	+	d	-
Lactose	-	-	-	-	-	-	-	-	-	+	-	+	-	-	-	-	d	-
Maltose	+	+	-	-	+	+	+	-	+	+	-	+	-	+	+	+	+	-
Sucrose	-	d	+	-	+	+	+	-	+	+	ND	+	-	+	+	+	d	-
Trehalose	-	-	-	-	d	-	d	d	+	+	d	-	-	+	+	+	d	-
Raffinose	d	-	-	-	-	-	-	-	-	-	ND	D	-	ND	-	-	-	-
Salicin	+	-	+	-	+	ND	-	-	-	-	ND	+	-	+	+	+	-	-
Dextrin	+	d	-	-	+	ND	+	+	+	+	-	+	-	ND	-	-	d	-
Starch	d	-	-	-	+	-	+	-	+	+	ND	-	-	-	-	-	-	-

ADAPTED FROM : Bergey's Manual of Systematic Bacteriology, (1986) Vol – 2, pp –1269

1. C. diphtheriae, 2. C. pseudotuberculosis, 3. C. xerosis, 4. C. pseudodiphtheriticum, 5. C. kutscheri, 6. C. minutissimum, 7. C. striatum, 8. C. renale, 9. C. cystitidis, 10. C. pilosum, 11. C. mycetoides, 12. C. matruchotii, 13. C. flavescens, 14. C. vitarumen, 15. C. glutamicum, 16. C. callunae, 17. C. bovis, 18. C. pannometabohm. ; + = 90% or more of strains are positive, - = 90% or more strains are negative, ND = Not defined. d = 11-89% of strains are positive.

117

TABLE - XXIV (Continued): CHARACTERISTICS DIFFERENTIATING THE SPECIES OF THE GENUS CORYNEBACTERIUM

Characteristics	1	2	3	4	5	6	7	8	9	10	11	12	13	14	15	16	17	18
HYDROLYSIS OF																		
Esculin	-	-	-	-	-	ND	-	-	-	-	-	+	ND	+	-	-	-	-
Hippurate	-	-	+	+	+	+	+	+	+	+	ND	+	-	-	+	+	+	-
Gelatin liquefaction	-	D	-	-	-	-	d*	-	-	-	-	d	-	-	-	-	-	-
Urease	-	+	-	+	+	+	-	+	+	+	-	-	-	+	+	+	-	+
Phosphatase	-	-	-	-	+	+	+	-	-	+	+	-	-	-	ND	ND	+	+
Decomposition of tyrosine	-	-	-	-	-	ND	+	-	-	-	ND	ND	ND	ND	ND	ND	+	+
Pyrazinamidase	-	-	+	+	-	+	ND	+	+	+	ND	+	-	+	ND	ND	+	+
Methyl red	+	+	-	-	-	-	+	-	-	-	-	-	+	+	-	+	-	-
Casein digestion	-	-	-	-	-	-	-	+	-	-	ND	ND	ND	ND	-	-	-	-
Nitrate - Nitrite	+	D	+	-	+	-	-	-	-	+	-	+	-	+	+	-	-	-
Tuberculostearic acid present	-	-	-	-	-	+	ND	-	ND	ND	-	-	-	-	-	-	+	+
Whip handle morphology	-	-	-	-	-	-	ND	-	-	-	-	+	-	-	-	-	-	-

ADAPTED FROM : Bergey's Manual of Systematic Bacteriology, (1986) Vol - 2, pp -1269

1. C. diphtheriae, 2. C. pseudotuberculosis, 3. C. xerosis, 4. C. pseudodiphtheriticum, 5. C. kutscheri, 6. C. minutissimum, 7. C. striatum, 8. C. renale, 9. C. cystitidis, 10. C. pilosum, 11. C. mycetoides, 12. C. matruchotii, 13. C. flavescens, 14. C. vitarumen, 15. C. glutamicum, 16. C. callunae, 17. C. bovis, 18. C. paurometabolum. ; + = 90% or more of strains are positive, - = 90% or more strains are negative, ND = Not defined. d = 11 -89% of strains are positive.

TABLE - XXV: Rf. VALUES OF MYCOLIC ACID AND OTHER LONG CHAIN COMPONENTS IN WHOLE – ORGANISM METHANOLYSATES OF 25 SUSPECTED CORYNEFORM BACTERIA

Strain PRL No.	No. Of Spots Visualized							
	1	2	3	1	1	2	3	4
	Rf – Value							
15	1.1	0.9	0.8	-	0.3	0.2	0.1	0.09
23	-	-	0.8	-	-	-	0.1	0.09
31	1.1	0.9	0.8	-	0.3	0.2	0.1	0.09
44	-	-	0.8	0.5	-	0.2	-	0.09
54	-	-	0.8	0.5	-	0.2	-	0.09
63	-	-	0.8	-	-	-	0.1	0.09
85	1.1	0.9	0.8	-	0.3	-	0.1	-
182	1.1	0.9	0.8	-	0.3	-	0.1	-
198	-	-	0.8	-	-	-	0.1	0.09
244	1.1	0.9	0.8	-	0.3	0.2	0.1	0.09
263	-	-	0.8	0.5	-	0.2	-	0.09
296	-	-	0.8	-	-	-	0.1	0.09
455	1.1	0.9	0.8	-	0.3	-	0.1	-
601	1.1	0.9	0.8	-	0.3	-	0.1	-
625	-	-	0.8	0.5	-	0.2	-	0.09
781	-	-	0.8	0.5	-	0.2	-	0.09
801	1.1	0.9	0.8	-	0.3	0.2	0.1	0.09
813	1.1	0.9	0.8	-	0.3	0.2	0.1	0.09
873	-	-	0.8	-	-	-	0.1	0.09
982	1.1	0.9	0.8	-	0.3	0.2	0.1	0.09
990	1.1	0.9	0.8	-	0.3	-	0.1	-
1139	-	-	0.8	-	-	-	0.1	0.09
1610	1.1	0.9	0.8	-	0.3	0.2	0.1	0.09
1990	1.1	0.9	0.8	-	0.3	-	0.1	-
2235	1.1	0.9	0.8	-	0.3	0.2	0.1	0.09
Identification	FAMEs			MAMEs	HYDROXY		FAMEs	

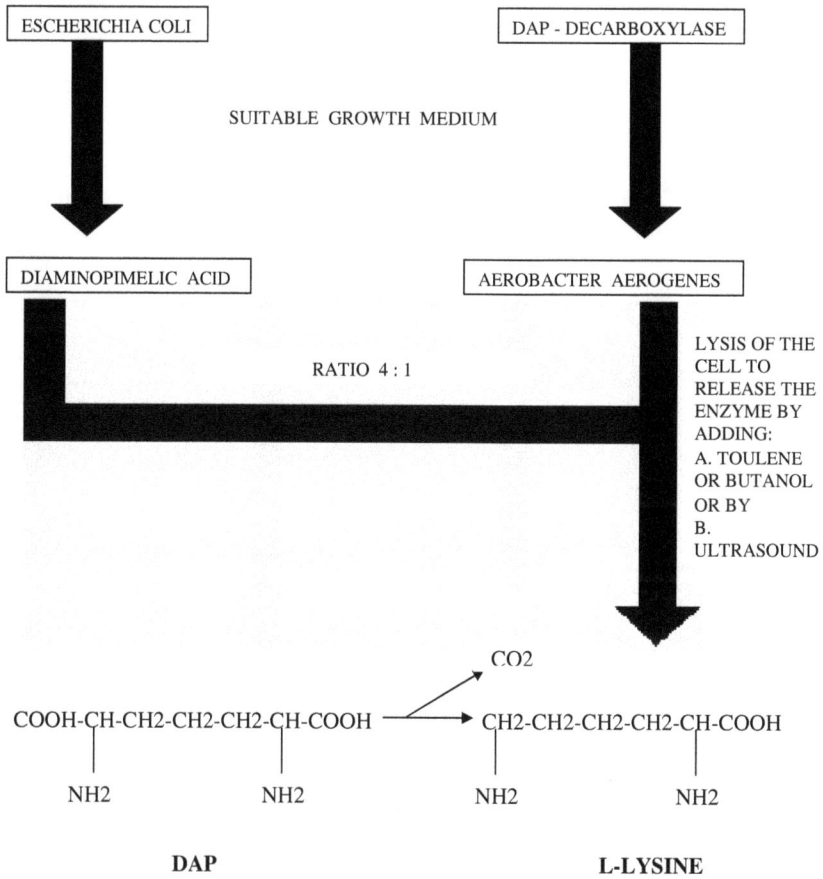

FIGURE – XIX: PRODUCTION OF L-LYSINE VIA CONVERSION OF ITS IMMEDIATE PRECURSOR – DAP (TWO – STEP PROCESS):

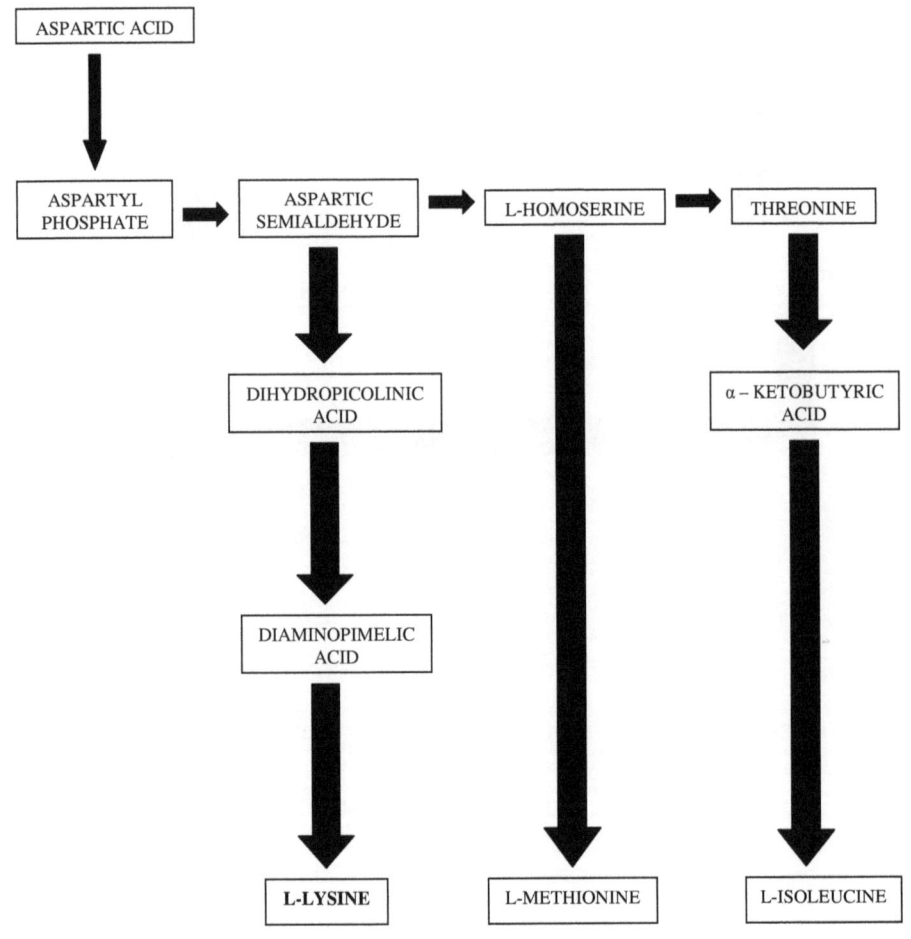

FIGURE – XX: PRODUCTION OF L-LYSINE BY THE AUXOTROPHIC MUTANTS:
(ONE STEP PROCESS)

ADAPTED FROM:

Nakayama, K.(1985). Lysine. In Comprehensive Biotechnology. Ed. Moo-Young, M. Vol.
3. Pergamon Press, Oxford,. pp. 607 - 620.

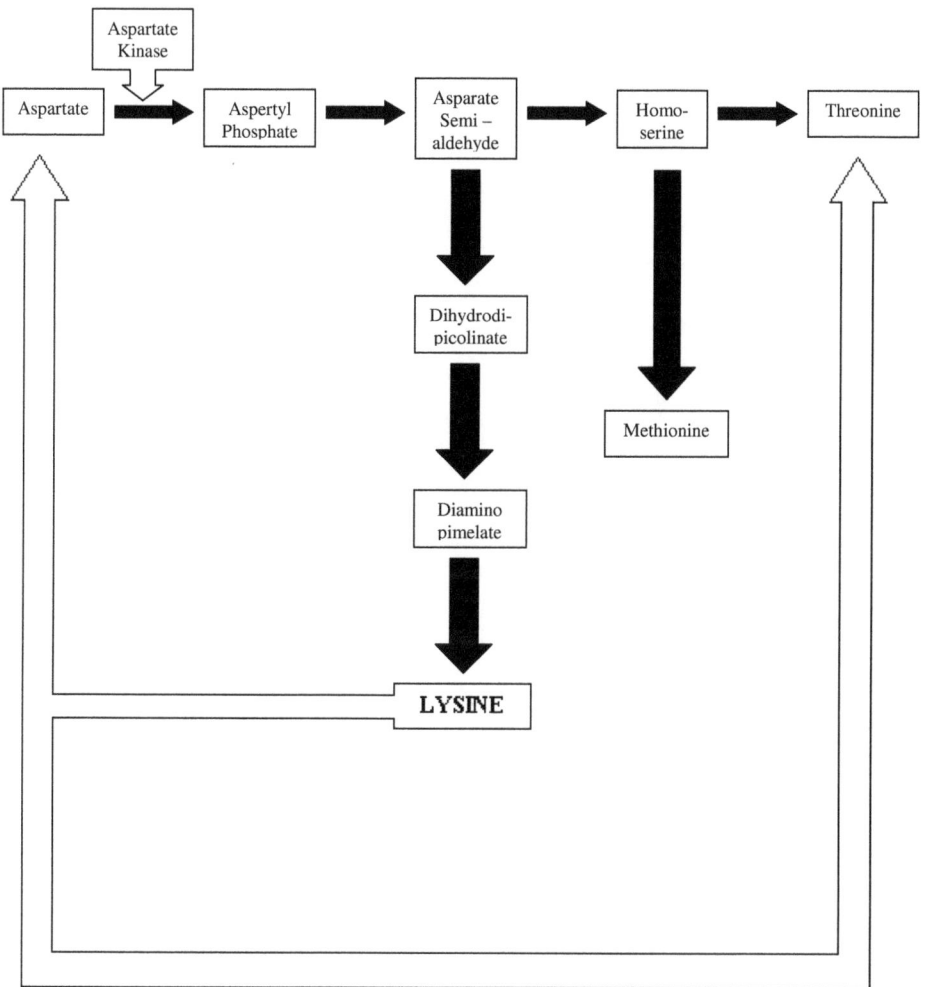

FIGURE – XXI: REGULATORY MECHANISM OF L-LYSINE BIOSYNTHETIC PATHWAY IN CORYNEACTERIUM GLUTAMICUM

MODIFIED FROM: Nakayama, K.(1985). Lysine. In Comprehensive Biotechnology. Ed. Moo-Young, M. Vol. 3. Pergamon Press, Oxford,. pp. 607 - 620.

- Large arrow indicates feedback inhibition.

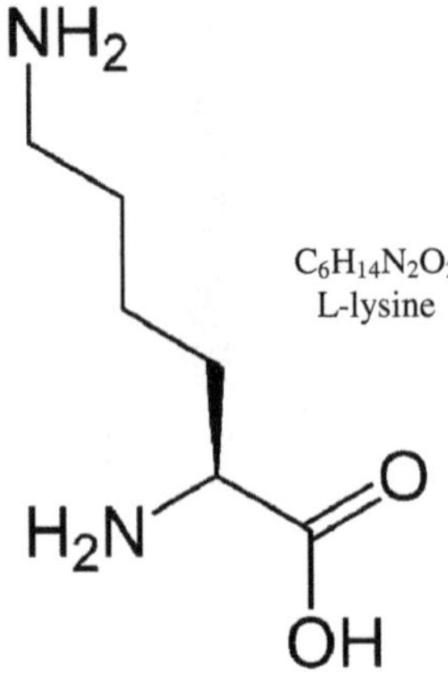

C6H14N2O2
L-lysine

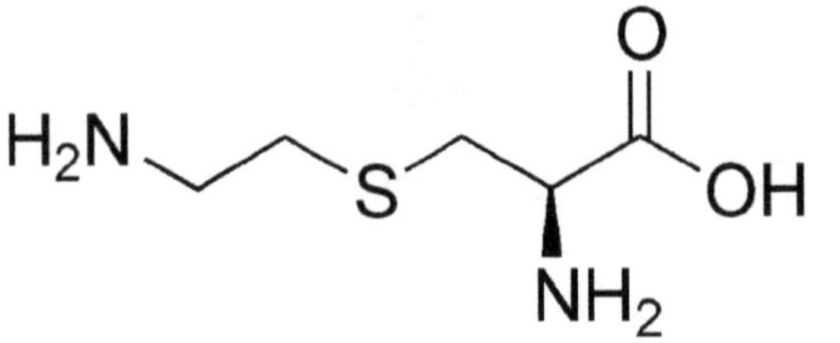

C5H12N2O2S
(2-Aminoethyl)-L-Cysteine
(AEC)

FIGURE - XXII : STRUCTURE OF L-LYSINE AND ITS ANALOGUE AEC

TABLE - XXVI: ANALYSIS OF THE COST OF RAW MATERIALS USED FOR THE PRODUCTION OF L-LYSINE

Raw Material	Price (US$/kg)	Amount used (g)		Total Cost (US$)	
		Medium Glucose	Medium Molasses	Medium Glucose	Medium Molasses
Glucose	0.23214	110.00	-	0.025536	-
Molasses	0.01935	-	275.00	-	0.005320
Yeast extract	2.32143	5.00	-	0.011607	-
Soyabean protein acid hydrolysate	0.09286	-	20.00	-	0.001857
Peptone	1.93452	10.00	-	0.019345	-
Cornsteep liquor	0.18571	-	5.00	-	0.000929
Fish meal	0.21667	5.00	5.00	0.001083	0.001083
Meat extract	1.16071	5.00	-	0.005804	-
Potassium di- hydrogen phosphate	1.47024	0.50	0.50	0.000735	0.000735
Di-potassium hydrogen phosphate	2.16667	1.50	1.50	0.003250	0.003250
Ammonium sulphate	0.12381	15.00	15.00	0.001857	0.001857
Calcium Carbonate	0.09286	10.00	10.00	0.000929	0.000929
Magnesium sulphate	0.07738	0.25	0.25	0.000019	0.000019
Manganese sulphate	0.07738	0.25	0.25	0.000019	0.000019
Ferrous sulphate	0.09286	0.01	0.01	0.000001	0.000001
Sodium acetate	0.37143	2.00	2.00	0.000743	0.000743
Vitamin B1	10.83333	0.001	0.001	0.000011	0.000011
Biotin	61.90476	0.00125	0.000025	0.000077	0.000002
Final Cost (US$)				**0.07102**	**0.01676**

Average amount of L-lysine produced (g/l) in medium   =   67.80   65.20

Cost of raw materials for the production of 1kg L-lysine(US$) =   1.0474   0.2569

1. The Price for each of the Raw Materials was based on the value obtained from local market
2. The amount used in the calculation was based on the laboratory scale experiment described in section 5.3 of materials and methods.

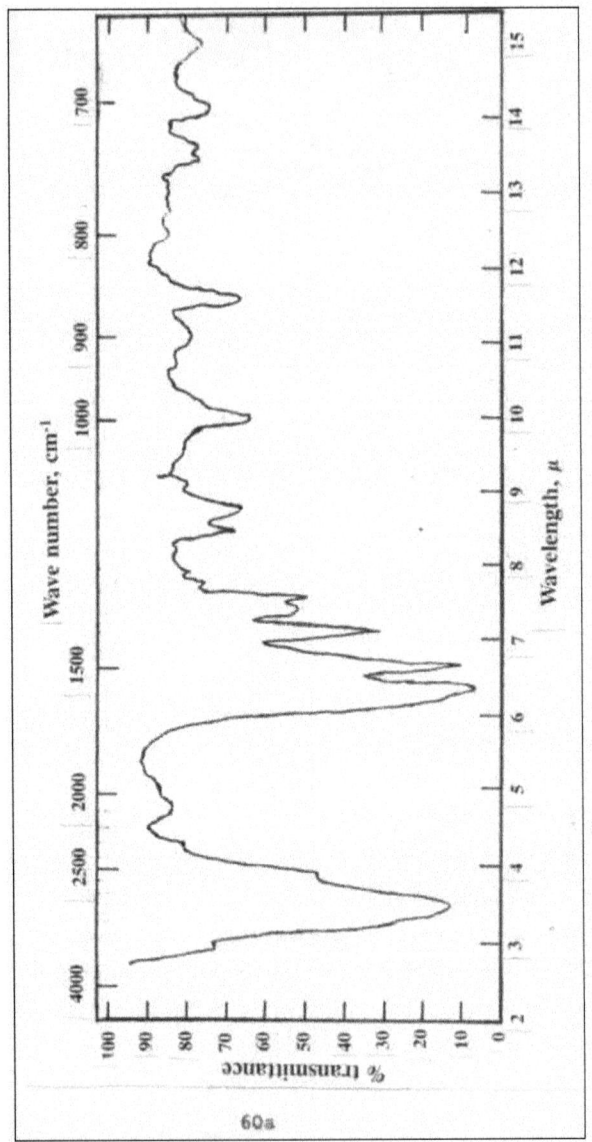

FIGURE –XXIII : A. INFRARED SPECTRUM OF STANDARD L-LYSINE

60a

125

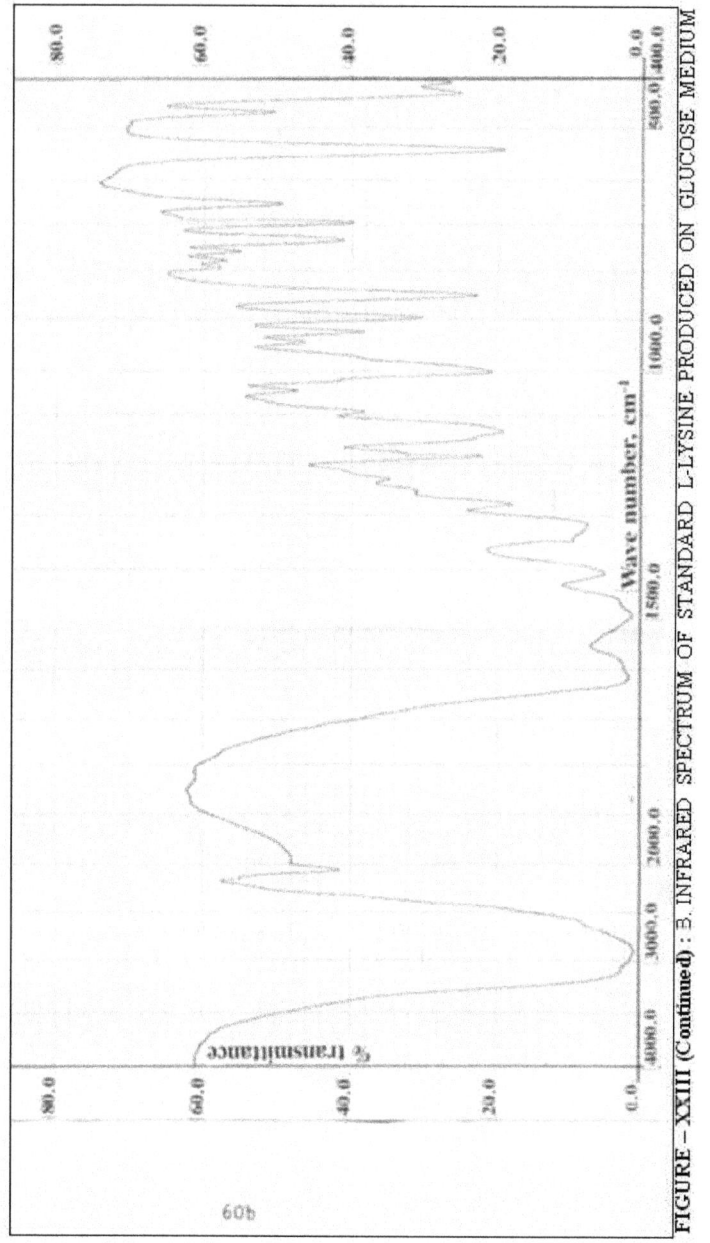

FIGURE – XXIII (Continued) : B. INFRARED SPECTRUM OF STANDARD L-LYSINE PRODUCED ON GLUCOSE MEDIUM

60b

126

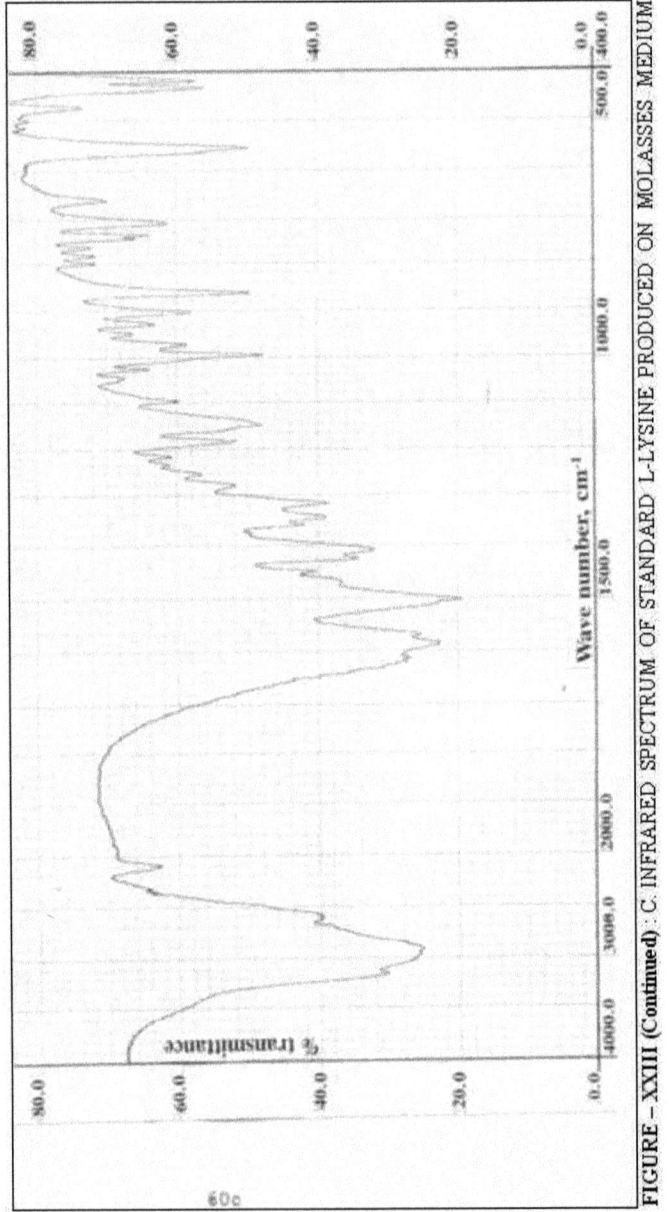

FIGURE – XXIII (Continued) : C. INFRARED SPECTRUM OF STANDARD L-LYSINE PRODUCED ON MOLASSES MEDIUM

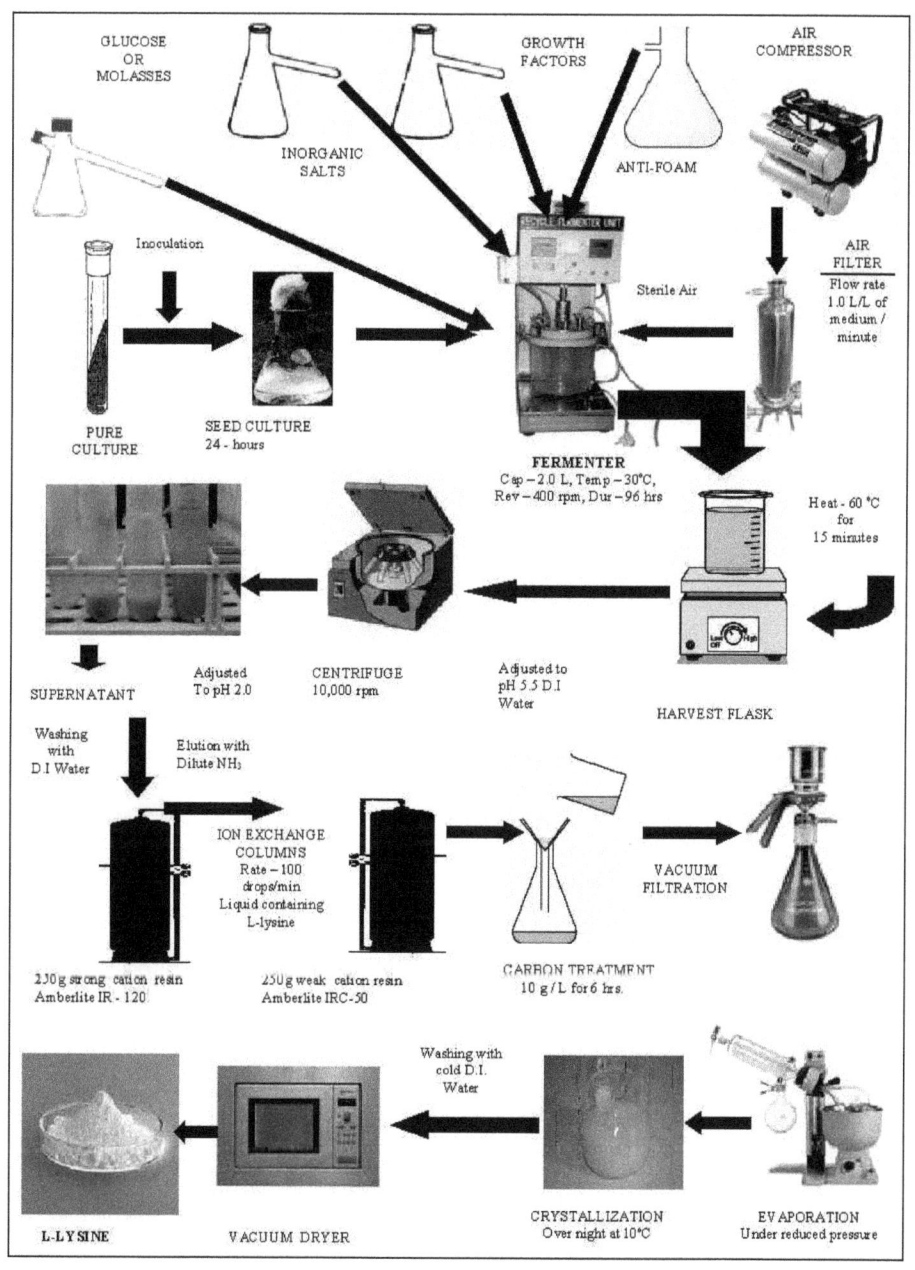

FIGURE – XXIV : LAY OUT OF L-LYSINE PRODUCTION BY FERMENTATION IN LABORATORY

ANNEXURE

ANNEXURE – I: NAME OF DISCOVERERS AND DATES OF AMINO ACIDS

AMINO ACID	DISCOVERER	YEAR
Alanine	Weyl	1888
	Schutzenberger	1879
Arginine	Hedin	1895
Asparagine	Demodaran	1932
Aspartic acid	Ritthausen	1868
Citruline	Koga	1914
	Odake	1914
	Wada	1930
Cystine	Morner	1899
	Emden	1899
Dihydroxyphenylalanine	Torquati	1913
	Guggenheim	1913
Glutamic acid	Ritthausen	1866
Glutamine	Damodaran, Jaaback, and Chibnall	1932
Glycine	Braconnot	1820
Histadine	Kossel	1896
Hydroxyproline	Fischer	1902
Isoleucine	Ehrlich	1904
Leucine	Braconnot	1820
Lysine	Drechsel	1889
Methionine	Mueller	1922
Ornithine	Riesser	1906
Phenylalanine	Schulze and Barbieri	1881
Proline	Fischer	1901
Serine	Cramer	1865
Threonine	Schryver and Buston	1925
	Gortner and Hoffmann	1925
Tyrosine	Bopp	1849
Tryptophan	Hopkins and Cole	1902
Valine	Fischer	1901

ADAPTED FROM: Considine, D.M. (1974).Chemical and process technology encyclopdia. McGraw-Hill Book Company, New York, pp – 95.

Gram staining:

1.**Method**: (A modified method of Hucker and Conn, 1923 and 1927).

Step-I: Preparation of smear: From 18 to 24 hours broth culture, one or two loopful of suspended cells were directly applied to a clean, dried glass slide and was allowed to dry. The slide was then fixed by passing it two or three times over the flame.

Step-II: Staining: The slide was flooded with ammonium oxalate - crystal violet solution and allowed to act for one minute, washed with tape water and flooded with Gram's iodine (mordant) to act for one minute. The slide was washed with tap water and decolourized with 95% ethyl alcohol, adding drop by drop until crystal violet fails to wash from smear. The slide was washed with tap water and counter stained with safranin for 45 seconds. It was again washed, dried in air and examined under oil immersion.

Results: Gram-positive organisms, blue; Gram-negative organisms, red in colour.

2. Acid-fast staining: Method: (Gross, 1952).

Step-I Preparation of smear: Materials and methods were same as described for Gram's staining.

Step-II Staining: The slide was flooded with carbol fuschin solution containing tergitol (sodium tetradecyl sulphate) and allowed to act for 5 minutes. The slide was washed with tap water and decolourized with acid-alcohol (HCl-ethyl alcohol), adding the reagent drop by drop until carbol fuschin fails to wash from smear. The slide was washed with tap water and counter stained with methylene blue for two minutes. After again washing with tap water the slide was blotted, dried in air and examined under oil immersion.

ANNEXURE - II (Continued)

Results: Acid-fast organisms, red; non-acid-fast organisms, appeared blue in colour.

3. Capsule staining: Method: (Hiss, 1905).

Step-I: Preparation of smear: Materials and methods were same as described for Gram's staining.

Step-II: Staining: The slide was flooded with crystal violet solution and gently heated for 5 to 7 minutes until steam rises. The slide was washed off with 20% aqueous $CuSO_4$. $5H_2O$, dried by blotting paper and examined under oil immersion.

Results: Capsule, cells dark purple, capsule colourless against a dark background.

4. Spore staining: Method: (Conklin, 1934).

Step-I: Prepration of smear: Metarials and methods were same as described for Gram's staining.

Step II: Staining: The slide was flooded wit 5% aqueous malachite green solution and steamed for 10 minutes, keeping slide flooded by adding fresh staining fluid. The slide was washed in running tap water for 30 seconds and counter-stained with 5% aqueous mercurochrome. The slide was washed in running tap water, dried by blotting paper and examined under oil immersion.

Results: Spores, green, rest of cells red.

5. Flagella staining: **Method** : (Bailey, 1929; Modified by Fisher and Conn, 1942).

Step-I Preparation of smear: 18 to 22 hours old cultures of the test organism (checked for motility) was used for the preparation of smear. The growth of the organism was washed off by gentle agitation with 2-3 ml of sterile distilled water and transferred to a sterile test tube. Following incubation at 35°C for 10 to 30 minutes, the culture was again checked for motility by hanging drop method. By means of a capillary pipette, a small drop of suspension from the top was transferred to one end of the slide. The slide was tilted and the drop was allowed to run slowly to the other end. The slide was kept in tilted position for some time to dry the smear in air.

Step-III Staining : The slide was flooded with solution-A (10% Tannic acid 18 ml and 6% $FeCl_2.6H_2O$ 6ml) and allowed to react for 3 to 4 minutes. The solution A was poured off and then without washing, the slide was flooded with solution-B (Basic fuchsin 0.5% in ethyl alcohol 0.5 ml, HCl concentrated 0.5 ml, and solution A 3.5 ml) and allowed to act for 7 minutes. The slide was washed with distilled water and covered with Ziehl's Carbol Fuchsin for one minute. After washing again with tap water, the slide was dried in air and examined under oil immersion.

Results: Flagella red, cells blue.

ANNEXURE – III: CULTURAL CHARACTERISTICS

(Cappuccino and Sherman, 1983).

1. Nutrient agar plates:

Method: A streak-plate inoculation of each test organism from 24 hours old broth culture was made. Plates were incubated at 35°C for 24 hours and colonies were examined for size, pigmentation, form, margin and elevation.

2. Nutrient agar slants:

Method: A single line streak inoculation of each test organism (24 hours broth culture) was made; starting at the butt and drawing the needle up the slanted agar surface. Following incubation at 35°C, for 24 to 48 hours, tubes were observed for growth, pigmentation, consistency and form.

3.Nutrient broth tubes:

Method: Nutrient broth tubes were inoculated with the test organisms (24 hours, old culture) and incubated at 35°C for 24 to 48 hours. The tubes were evaluated for the distribution and appearance of the growth.

4. Nutrient agar stab:

Method: Test organisms (24 hours old nutrient broth culture) were inoculated in nutrient agar tubes by means of a stab inoculation and incubated at 35°C for 24 to 48 hours. At the end of incubation period tubes were examined for pattern of growth along the stab line. (i.e. motile or non-motile). If negative, tubes were further incubated for 5 days at 25°C.

ANNEXURE –IV: BIOCHEMICAL CHARACTERISTICS

(MacFaddin, 1976; Cappuccino and Sherman, 1983).

1. Carbohydrate fermentation test:

Method: A series of test tubes containing sterile phenol red broth with single fermentable carbohydrate (1%) and Durham's tubes were inoculated with test organisms grown on TSB medium for 24 hours. Sugar, tubes were incubated at 35°C for 24 hours. At the end of incubation period, all tubes were examined for production of acid and gas and compared with control. Tubes showing negative result were kept under prolong incubation (upto 30 days) for further confirmation.

Results: A change in colour of broth from red to yellow indicated acid production, while presence of bubble in Durham's tube indicated formation of gas. Production of acid or acid with gas indicated a positive carbohydrate fermentation test.

2. Catalase test:

Method: With the help of an inoculation needle, the center of an 24 hours culture of the test organisms (grown on nutrient agar) were picked and placed on clean glass slides. A drop of 30% H_2O_2 was added over the culture on the slides. The slides were immediately observed for bubbling (gas liberation) and compared with the control.

Results: Immediate bubbling (O_2 formation) indicated a positive test.

3. Coagulase test:

Methation-I: Slide test (Presumptive test): A drop of sterile distilled water was placed on clean glass slides. A loopful of 24 hours broth culture of the test organisms were gently emulsified with a droop of pretested rabbit plasma. The slides were immediately observed for the formation of macroscopic precipitates in the form of white clumps and compared with the control.

Method-II: Tube test (Confirmative test): Pretested rabbit plasma (0.5 ml) was added to a sterile glass test tubes. To each tube 0.5 ml of the 24 hours old culture of the test organisms were added. The tubes were gently rotated and incubated at 35°C for 4 hours. At the end of incubation period, the tubes were observed for clotting and compared with the control.

Result: A marked clumping within 5 to 20 seconds in the slide test indicated a positive test, whereas any degree of clumping or granulation after 20 seconds and upto 1 minute, indicated a delayed positive test; while any granulation after 1 min., indicated a doubtful test. While formation of a clot or distinct fibrin threads throughout the tube indicated a complete positive test, whereas any degree of clot extended throughout the tube indicated a partial positive test.

4. Decarboxylase test: (Lysine-Ornithine-Arginine).

Method: Test tubes of decarboxylase broth containing lysine, ornithine and arginine in a concentration of 1% were marked A,B and C respectively. The tubes were inoculated from a 24 hours old culture of the test organism, including control, containing no amino acid. To each tube 1 to 2 ml sterile mineral water was added and tubes were incubated at 35°C for 24 hours. At the end of incubation period, the tubes were observed for colour development.

Results: Development of purple colour indicated a positive test.

5. Gelatin liquefaction test:

Method: Test organims (24 hours old broth culture grown in TSB) were inoculated in the deep nutrient gelatin tubes by means of a stab inocultaion. The tubes were incubated at 35°C for 24 hours. At the end of incubation period, the tubes were placed in refrigerator at 4°C for 30 minutes and observed for the liquefication of medium and compared with the control.

Results: Liquefication of gelatin incidated a positive test.

6. Hydrogen sulphide test:

Method: Test organisms (24 hours broth culture grown in TSB) were inoculated into deep TSB agar tubes, by means of stab inoculation. The tubes were inoculated at 35°C for 48 hours. At the end of incubation period, tubes were observed for the formation of hydrogen sulphide and compared with the control.

Results: The presence of black colouration throughout the entire butt or along with the stab inoculateion, indicated a positive test (formation of hydrogen sulphide).

7. Indole test:

Method: The test tubes containing sterile tryptone broth were inoculated with a loopful of 24 hours old broth cultures grown in TSB of the test organisms. Tubes were incubated at 35°C for 24 to 48 hours. At the end of incubation period, 5 drop of Kovac's reagant was added directly to the tubes, shaken gently and observed for the production of indole and compared with the control.

Results: A red ring at the surface of the medium indicated a positive result.

8. Litmus milk test:

Method: Test tubes containing sterile litmus milk broth were inoculated with the 24 hours old broth culture (grown in TSB) of the test organisms by means of a loop inoculation. Following incubation at 35°C for 24 hours, the tubes were examined for the change in colour and consistency and compared with the control.

Results:

I. A change of colour from purple to pink or pinkish-red (acid reaction lactose fermentation) indicated a positive test.

II. Production of clot or curd (milk protein coagulation) indicated a positive red.

III. Reduction of limits to white or milk coloured (litmus reduction) indciated a positive test.

IV.The separation of the curd, or the development of tracks or fissures within the curd (evolution of gases) indicated a positive test.

9. Methyl red test:

Method: Test tubes containing sterile MR/VP broth were inoculated from a 24 hours old broth culture (grown in TSB) of the test organisms by means of a loop inoculatuion. The tubes were incubated at 30°C for 3 to 5 days. At the end of incubation period, methyl red indicator was directly added to each tube and observed for change in colour and compared with the control.

Results: Maintenance of a distinct red colour at the surface of the medium, indicated a positive test.

10. Nitrate reduction test:

Method: Test tubes containing sterile trypticase nitrate broth were inoculated from a 24 hours old broth culture, (grown in TSB) by means of a loop inoculation technique. The tubes were incubated at 35°C for 24 hours, and at the end of incubation period nitrate reagents were added. The tubes were immediately observed for the development of colour and compared with the control tube.

Results:

Phase-I: Development of red colour by the addition of five drops of sulfanilic acid fol
          lowed by five drops of dimethyl alpha-naphthylamine indicated a positive test.

Pase-II: A minute quantity of zinc powder was added to those tubes in which there has
          been no red colour development in the Phase-I. Development of no colour,
          indicated a positive test.

11. Oxidase test:

Method: Tryticase soy agar plates were inoculated by means of streak plate technique from a 24 hours old nutrient broth culture of the test organisms. The plates were incubated in an inverted position at 35°C for 24 hours. At the end of incubation period, Kovac's reagent (two to three drops) was added to the surface of the growth and observed for the presence or absence of a colour change from pink, to maroon and finally to black.

Results: Change of colour from pink to maroon and finally to black, indicated a positive oxidase test.

12. Phosphatase test:

Method: Test tubes containing sterile phenolphthalein diphosphate medium   (PDP-broth) were inoculated with the 4 hours old broth culture (grown in TSB) of the test organisms. The tubes were incubated at 35°C for 6 hours (If negative, the second tube was incubated for 24 hours). At the end of incubation period, 1 drop of 10 N NaOH was added to the tubes and observed for the change in colour and compared with the control.

Results: A change in colour from white to bright pink-red colour, indicated a positive test.

13. Urase test:

Method: Test tube containing sterile urea broth was inoculated with the 24 hours old culture (grown in TSB) of the test organisms. Following incubation at 35°C for 48 hours, the tubes were observed for the development pink-red colour and compared with the control.

Results: A change of colour from yellow-orange to pink-red indicated a positive test.

14. Voges-Proskauer test:

Method: Test tubes containing sterile MR-VP broth media were inoculated with the 24 hours old culture (grown in TSB) of the test organisms. Following incubation at 35°C for 48 hours, the Voges-Proskauer reagant was directly added to the tubes and observed for a change in colour and compared with the control.

Results: Development of pinkish-red colour at the surface of the medium indicated a positive test.

15. Starch hydrolysis test:

Method: Starch agar plates were streaked with the 24 hours old culture (grown in TSB) of the test organisms. Following incubation in an inverted position for 48 hours at 35°C, Gram's iodine solution was added directly to the surface of the medium for 30 seconds. After   pouring   off, the plates were observed for the presence of blue -black   colour surrounding the growth and compared with the control.

Results: Presence of a blue-black colour surrounding the growth indicated (starch hydrolysis) a positive test.

16. Casein hydrolysis test:

Method: Milk agar plates were inoculated with the 2 hours old culture (grown in TSB) of the test organisms. Following incubation at 36°C for 48 hours, the plates were observed for the presence of clear zone and compared with the control.

Results: Presence of a clear zone of proteolysis indicated a positive test.

ANNEXURE – VI: PHYSIOLOGICAL CHARACTERISTICS

(Cappuccino and Sherman, 1983)

1.Temperature:

Method: The test organisms were aseptically inoculated in four nutrient agar slants and incubated   for 24 to 48 hours at 4°C, 20°C, 40°C and 60°C. At the end of incubation period, all tubes were observed for the presence of growth at different temperature.

2.pH:

Method: Using   sterile   pippetes, a series of trypticase soy broth tubes (pH values of 3,5,7 and 9) were inoculated with 0.1 ml of the saline culture of the test organisms and incubated at 35°C for 48 hours to 72 hours. At the end of incubation period, all tubes were observed for the presence of growth, at different pH values.

3. Oxygen:

Method: The test organisms were inoculated to molten nutrient agar tubes (at 45°C), shaken and immediately placed into the ice-water bath in an upright position to solidify the medium   rapidly.   Tubes were incubated at 35°C for 24 to 48 hours. At   the   end   of incubation period, tubes were examined for the distribution of growth.

 4. Heat:

Method: The test organisms (24hours old nutrient broth culture) were given heat shock at 40°C, 60°C, 80°C and 100°C for 10 minutes and were inoculated in their appropriate nutrient agar plates labeled as 40°C, 60°C, 80°C and 100°C. Plates were incubated at 35°C for 24 to 48 hours. At the end of incubation period, all plates were observed for the amount of growth at different temperature.

5. NaCl:

Method: The test organisms (24 hours old nutrient broth culture) were inoculated into 5 different nutrient agar plates, each containing 0.85%, 5%, 10%, 15% and 25% sodium chloride. Plates were incubated at 35°C for 4 to 5 days. At the end of incubation period all plates   were observed for the amount of growth at different concentration   of   sodium chloride.

ANNEXURE - VI (A): CHEMICAL CHARACTERISTICS:

1.   **Principal   amino acids and amino sugars in the cell wall**: (Yamada   and Komagate, 1970a).

Preparation of cell wall: Bacteria to be tested were grown in a   liquid   medium composed of peptone, 10 g; yeast extract, 5 g; casamino acid; 5 g; Tween 80, 0.05 g; distilled water, 1000 ml; and adjusted to pH 7.2, for 20 to 48 hours at 30°C with shaking. For the poor growth strains, 2 g of brain-heart infusion was added to the medium. Bacterial cells in logarithmic phase were harvested by centrifugation at low temperature and washed twice with distilled water. Cells were disrsupted by mechanical cell homogenizer. About 10 g of wet packed cells suspended in an equal volume of distilled water, 20g of glass beads (0.25-0.30 mm) and 0.5 g polypropylene glycol (anti-foaming agent) were mixed in a 75 ml glass flask. After sufficient ice-cooling, the cells were disintegrated at 4000 rev/min until disruption of more then 95% was attained. The degree of disruption was confirmed by Gram-stain. The suspension was separated from glass beads through a sintered glass filter

137

and the beads were washed with minimal amount of distilled water. Further, unborken cells were excluded by centrifugation twice at 1000 rev/min for 10 minutes. A fraction of cell wall was obtained by further centrifugation at 19000 rev/min for 10 minutes and washed twice with distilled water. The presence of unbroken cells was checked by microscopy.

The final preparation was resuspended in 0.05 M phosphate buffer (pH 7.6) and digested by 0.5 mg/ml of trypsin and 0.5 mg/ml of ribonuclease at 37°C for 5 hours. The residue was washed twice with distilled water, suspended in 0.002 N HCl, and digested by 1 mg/ml of crystalline pepsin at 37°C for 24 hours. After this treatment the residual material was washed twice with distilled water, followed successively by 30%, 50%, 80%, and 99% ethanol. The residue dried in vacuum was regarded as the cell wall.

Hydrolysis of cell wall and paper chromatography of amino acids and amino sugars:

About 50 mg of cell material was hydrolyzed with 6 N HCl in a sealed ampule at 105°C for 8 hours. The hydrolyzate was filtered and evaporated to dryness in vacuum. The dry matter was dissolved in a small amount of distilled water and subjected to paper chromatography. The following three solvent systems were used: (1) Phenol saturated with water plus 0.1% ammonia for the detection of amino acids; (2) lutidine saturated with water for the detection of amino sugars; and (3) methanol-water-6N HCl-pyridine (80:26:4:10 v/v) for the separation of stereoisomers of diaminopimelic acid (DAP). Aluminnium T.L.C plates of 0.2 mm thickness were employed throughout this work. Amino acids and two isomers of DAP contained in bacterial cell wall were detected on one-dimensional descending chromatogram. Since the isomers of DL- and DD-DAP were not separated by the paper chromatography and DD-DAP seemed to be very uncommon in the microbial cell walls, the spot corresponding to DL or DD-DAP was recorded as DL-DAP.

ANNEXURE - VI (B): CHEMICAL CHARACTERISTICS

2. **Mycolic acid and other long-chain components in whole-organism methanolysates**: (Minnikin et al. 1975).

Cultivation of test organisms: The bacteria were grown in nutrient broth supplemented with 0.05 g tween 80 per 1000 ml of medium (pH 7.2) for 72 hours at 30°C on a rotary shaker at 250 rpm. After cultivation, bacteria were harvested by centrifugation at 7000 rpm, washed with distilled water and dried in a vacuum desiccator over $CaCl_2$ to a constant weight.

Whole-cell methanolysis: Dry bacteria (approx. 100 mg) were mixed with dry methanol (5 ml), toluene (5 ml) and concentrated sulphuric acid (0.2 ml) in a 20 ml capped tube. The contents of the tube were mixed throughly and methanolysis was allowed to proceed for 16 hours at 50°C. Initially,tubes were shaken vigorously in a water bath. The reaction mixture was allowed to cool at room temperature, 2 ml hexane was added, the mixture was shaken and then allowed to settle.

Thin-layer chromatography: Samples from the upper layer of the mixture were spotted on thin-layer plates (Aluminium T.L.C plates of 0.2 mm thickness, DC-Mikrokarten SIF 20 X 20 cm, Riedel-De-Haen Aktiengesellschaft Seelze-Hannover). The chromatograms were developed in petroleum ether (b.p 60 to 80°C) - diethyl ether (85:15, v/v). The position of the separated components were revealed by charring at 150 to 200°C after spraying with chromic acid solution (5 g $K_2Cr_2O_7$ in 5 ml water, made up to 100 ml with conc. H_2SO_4, then diluted ten times with water). The mycolic esters were distinguished from fatty acids of lower molecular weight by reversibly detecting the spots with iodine vapour, and then redeveloping the chromatogram in methanol - water (5:2, v/v) for 6 hours.

ANNEXURE - VI (C) : CHEMICAL CHARACTERISTICS

3.   DNA base composition:

Cultivation of test organisms: (Abe et al. 1967).

A 50 ml of seed culture (of each test organism) grown for 20 hours at 30°C in nutrient broth was inoculated into 500 ml of glucose (1%) bouillon medium contained in a 2.0 l Erlenmeyer flask and cultured at 30°C for 6 hours on a rotary shaker. Bacterial cells were harvested by centrifugation at a low temperature (7000 rpm at 4°C), washed twice with 0.15 M NaCl plus 0.1 M EDTA, pH 8.0 and dried in a vacuum desiccator over $CaCl_2$ to a constant weight.

DNA preparation: (Keleti and Lederer, 1974)

30 g dried bacterial cells were suspended in 500 ml EDTA (0.1 M) and 500 ml saline (0.15 M NaCl) at pH 8.0. The material was mixed and kept in a water bath maintained at 20°C. After 24 hours 13.11 g sodium dodecyl sulphate was added and the mixture was

stirred for 5 hours at 20°C in a water bath. Sufficient quantity of 95% ethanol was added until a flocculant precipitate was formed. The mixture was centrifuged at 10,000 rpm for 30 minutes at 4°C. The precipitate was washed twice with 70% ethanol and centrifuged at 6000 rpm for 25 minutes at 4°C. The precipitate was dissolved by vigorous stirring in 600 ml NaCl (1.5 M) in a blender. The material was centrifuged at 15,000 rpm for 1 hour at 4°C and the pellet was discarded. To the supernatants 500 ml chloroform:octanol (8:1) was added and placed in a separating funnel, shaked and the lower organic phase was dicarded. This step was repeated twice. The saline solution was precipitated with isopropyl alcohol and centrifuged at 10,000 rpm for 60 mintues at 4°C, washed twice with 70% isopropyl alcohol and centrifuged at 6000 rpm for 25 minutes at 4°C. The precipitate was dissolved in saline (0.15 M NaCl) and 20 mg ribonuclease was added. The material was placed in a water bath maintained at 37°C for 1 hour and shaken intermittently. An equal volume of 90% phenol was added to the material, placed in a water bath maintained at 20°C and stirred vigorously for 20 minutes. The material was cooled in an ice bath and centrifuged at 7000 rpm for 45 minutes at 4°C. The upper water layer containing the DNA was aspirated. An equal volume of saline to the phenol layer was added and stirred vigorously in a water bath at 20°C for 20 minutes. The mixture was cooled in an ice bath, centrifuged at 7000 rpm for 45 minutes at 4°C and the upper water layer containing DNA was aspirated. The aqueous extracts were combined and dialyzed against water for three days to remove the phenol. It was precipitated with 70% isopropyl alcohol and centrifuged at 15,000 rpm for 30 minutes at 4°C. The supernatant was discarded and the precipitate was washed twice with 70% isopropyl alcohol and centrifuged at 10,000 rpm for 30 minutes at 4°C. The precipitate was dissolved in saline (0.15 M NaCl) and centrifuged at 45,000 rpm for 2 hours. The pellet was discarded and recentrifuged at 65,000 rpm for 2 hours at 4°C. The pellet was one again discarded. The supernatant contains the purified DNA.

Determination of DNA base composition: (Marmur and Doty, 1962; Deley and Schell, 1963; Yamada and Komagata, 1970b).

DNA based composition was calculated from its thermal denaturation temperature (Tm). A Perkin-Elmer Spectrophotometer (Hitachi Ltd., Tokyo) equipped with thermospacers for circulating ethylene glycol was used. The DNA solution were heated continuously by means of a temparture programme controller adjusted to give a rate of increasing that does not exceed 0.25 °C / min. Tm in the standard saline-citrate solution (O.15 M Nacl plus 0.015 M tri-sodium citrate) was determined with a thermister of which thermosensor was inserted directly in DNA solution. The thermal denaturation (Tm) was followed at 260 nm and determined graphically. The DNA base compositions were calculated from the melting temperatures according to the following equations:

$\% \ G + C = 2.44 \ Tm - 169.3$ (DeLey, 1970).

ANNEXURE – VII: CLASSIFICATION OF THE PROCESSES FOR AMINO ACID PRODUCTION ACCORDING TO THE TYPE OF MICROORGANISM EMPLOYED

Type of Microrganism	Amino acid Produced	Representative example	
		Important Characters of the microrganism	Microorganism
I. Wild type	L-Glutamic acid	Requires biotin	*Corynebacterium glutamicum*
	L-Glutamine	Requires biotin	*Corynebacterium glutamicum*
	DL-Alanine		*Corynebacterium glutamicum*
	L-Valine	AL synthetase insensitive of valine	*Paracolobactrum coliforme*
II. Auxotrophic mutants	L-Lysine	Requires homoserine	*Corynebacterium glutamicum*
	L-Ornithine	Requires arginine	*Corynebacterium glutamicum*
	L-Threonine	Requires methionine, isoleucine and diaminopimelic acid	*Escherichia coli.*
III. Regulatory mutants	L-Lysine	Resistant to AEC	*Corynebacterium glutamicum*
	L-Isoleucine	Resistant to $\alpha – AB$	*Serratia marcescens*
	L-Arginine	Resistant to D-arginine and arginine hydroxamate	*Corynebacterium glutamicum*
	L-Histidine	Resistant to TRA	*Corynebacterium glutamicum*
IV. Auxotrophic regulatory mutants	L-Threonine	Resistant AHV and Thialysine and requires methionine.	*Corynebacterium glutamicum*
	L-Tyrosine	Resistant to aromatic aminoacid Analogs and requires phenyl-alanine.	*Corynebacterium glutamicum*
	L-Leucine	Resistant to AHV and requires isoleucine.	*Serratia marcescens.*

AL = Acetolactate, AEC = S-(β - aminoethyl)- L-cysteine, $\alpha – AB = \alpha$ - aminobutyric acid
TRA = 1,2,4 – triazolealanine, AHV = $\alpha – amino - \beta$ - hydroxyvaleric acid

ADAPTED FROM: Nakayama, K. (1972b). Microorganisms in amino acid fermentation. *Proc. IV IFS: Ferment. Technol. Today*, pp. 433.

REFERENCES

Abe, S., Takayama, K. and Kinoshita, S. (1967). Taxonomical studies on glutamic acid producing bacteria, *J. Gen. Appl. Microbiol.* **13**, 279.

Adelberg, E. A. and Myers, J.W. (1953), Modification of the pencillin technique for the selection of auxotrophic bacteria. *J. Bacteriol.* **65**, 348.

Akashi, K., Shibai, H. and Hirose, Y. (1979), Effect of oxygen supply on L-lysine, L threonine and L-isoleucine fermentations. *Agric. Boil. Chem.* **43** (10), 2087.

Alfredo, S.M., Lilia, V., Sara, M. and Hiranaka, H. (1969), Extracellular production of L lysine by mutants of *Ustilago maydis* in agave juice. *Rev. Lationamer. Microbiol. Parastiol.* **11** (4), 183., Chem. Abs **72:**109793y (1970).

Anderson, R.F. and Jackson, R.W. (1958). Essential amino acids in microbial proteins *Appl. Microbiol.* **6**, 369.

Angulo, J., Maurinio, T.D. G., Herrera. J., Municio. A.M. and Rivero, W. (1960a). Biosynthesis of a, e -diaminopimelic acid via *Escherichia coli*, IV. Utilization of Polyalcohols.

Anales Real Sco, Espan. Fis Y Quim **56B**, 311., Chem. Abs. **55**: 3706i, (1961).

Anastassiadis, S. (2007). L-lysine fermentation. *Recent Patents on Biotechnology*, 1, 11 –24.

Angulo, J., Maurinio T.D.G., Herrera. J., Municio, A.M. and Rivero, W. (1960b). Biosynthesis of α ε -diaminopimelic acid by *Escherichia coli*. VI Growth of a lysine auxotrophic mutant. *Anales Real Soc. Espan. Fis Y. Quim* **56B**, 431., Chem. Abs. **55**: 9562b (1961).

Areshkina, L.Y., Bukin, V.N., Bekeris, V., Bekeris, M., Valdmanis, A., Karklins, R., Kutseva., L.S., Klyueva, N.M., Liepins G. and Ramina, L. (1965a). Production of L-lysine. *U.S.S.R. Patent* 171,727., Chem. Abs **63**: 15509g (1965).

Areshkina, L.Y., Beker, M.E., Bukin, V.N., Karklins, R., Klyueva, N.M., Kutseva, L.S. and Liepims, G. (1965b). Microbiological synthesis of L-lysine. *Prikl. Biokhim. i Mikrobiol.* **1**(4). 396., Chem. Abs. **64**: 1315f(1966).

Ariga, N., Maruyama, K. and Kawaguchi. A. (1984). Comparative studies of fatty acid syntheses of Corynebacteria *J. Gen. Appl. Microbiol.* **30**, 87.

Asai, T., Okumura, S. and Tsunoda, T. (1957). On the classification of the a Ketoglutaric acid accumulating bacteria in aerobic fermentation. *J. Gen. Appl. Microbiol.* **3**(1), 13.

Baile, H.D. (1929), A. flagella and capsule stain for bacteria. *Proc. Soc. Exptl. Biol Med.* **27**, 111.

Balch, J.F. and Balch, P.A. (1990). Prescription for nutritional healing. Avery Publishing Group Inc, New York, p. 27.

Barksdale, L. (1970), *Corynebacterium diphtheriae* and its relatives. *Bacteriol, Rev.* **34**,378.

Barnes, I.J., Bondi, A. and Moat, A.G. (1969). Biochemical characterization of lysine auxotrophs of *Staphylococcus aureus*. *J. Bacteriol.* **99**(1), 169.

Bartlett, A.T.M. and White, P.J. (1985). Species of Bacillus that make a vegetative peptidobylcan containing lysine lack diaminopimelate epimerase but have diaminopimelate dehydrogenase. *J. Gen. Microbiol.,* **131**, 2145 – 2152.

Barton-Wright, E.C. (1952). The microbiological assay of the vitamin B-complex and amino acids. Sir Isaac Pitman and Sons, Ltd. London, p. 154.

Becker, J., Klopprogge, C., Schröder, H. and Christoph Wittmann. (2009). TCA cycle engineering for improved lysine production in *Corynebacterium glutamicum. Appl. Environ. Microbiol.* **75**(24). 7866 – 7869.

Bergey's Manual of Determinative Bacteriology. (1994)., Eds John G. Holt et al., 9th ed. The Williams and Wilkins, Baltimore, p. 565.

Bergey's Manual of Systematic Bacteriology. (1986). Ed. Sneath, P.H.A. vol. 2. Williams and Wilkins, Baltimore, p. 1261.

Bhattacharjee, J.K. and Sinha, A.K. (1972). Relations among the genes, enzymes, and intermediates of the biosynthetic pathway of lysine in *Saccharomyces. Mol. Gen. Genet.* **115**(1), 26.

Birge, E.A. (1988). Bacterial and bacteriophage genetics, 2nd ed., Springer-Verlag, New York, p. 58.

Blombach, B., Hans, S., Bathe, B. and Eikmanns, B.J. (2009). Acetohydroxy synthase, a novel target for improvement of L-lysine production by Corynebacterium glutamicum, *Appl. Environ. Microbiol.* **75**(2), 419 – 427.

Bopp, F. (1849). Einiges über albumin, casein and fibrin (Eng: Studies on albumin, casein and fibrin). *Ann* **69**,16.

Born, T.L. and Blanchard, J.S. (1999). Structure / function studies on enzymes in the diaminopimelate pathway of bacterial cell wall biosynthesis. *Curr. Opin. Chem. Boil.* **3**, 607 – 613.

Braconnot, H. (1820). Sur la conversion des matieres animales en nouvelles substances par le moyen de l'acid sulfurique (Eng: On the conversion of animal materials to new substances by means of sulphuric acid). *Ann. Chim, Phys.* **13**(2), 113.

Brenner, M., Niedewieser, A. and pataki, G. (1969). Amino acids and derivatives. In Thin Layer Chroatography. Ed. Stahl, E. 2nd ed. Springer-Verlag, Berlin, P. 730.

Broeer, S., Eggeling, L. and Kraemer, R. (1993). Strains of *Corynebacterium glutamicum* with different lysine productivities may have different lysine excretion systems. *Appl. Environ. microbiol.* **59**(1),316.

Broeer, S. and Kraemer, R. (1991), lysine exceretion by *Corynebactrium glutamicum.* I. Identificaiton of a specific secretion carrier system. *Eur. J. biochem.* **202**(1),131.

Broquist, H.P. (1982). Carnitine biosynthesis and function. Introductory remarks. *Fed. Proc.* **41**, 2840 – 2842.

Bucko, M., Hano, A., Scipakova, J., Salka, J. and Varga, V. (1989), L-lysine production with *Micrococcus glutamicus* auxotrophs. *Czech. CS.* 260,706., Chem. Abs. **112**: 117343m (1990).

Bulfer, S.L., Scott, E.M., Pillus, L. and Trievel. R.C. (2010) Structural basis for L-lysine feedback inhibition of homocitrate sysnthase. *J. Biol. Chem.*

Cappuccino, J.G. and Sherman, N. (1983), Microbiology: A laboratory manual. Addison Wesley Publishing Company, Massachusetts, p. 87.

Casida, L.E. and Baldwin, N.Y.(1956). Preparation of diaminopimelic acid and lysine *U.S. Patent* 2,771,396.

Cassan. M., Boy, E., Borne, F., and Patte, J.C. (1975). Regulation of the lysine biosynthetic pathway in *Escherichia coli* K-12. Isolation of a cis-dominant constitutive mutant for AK (aspartokinase) III synthesis. *J. Bacteriol* **123**(2), 391,

Chaitow, L. (1985). Amino acids in therapy. A guide to the therapeutic application of protein constituents. Thorsons Publishers Limited, Northamptonshire, p.52.

Champe, P.C. and Harvey, R.A. (1993). Lippincott's Illustrated Reviews: Biochemistry. 2nd ed., J.B. Lippincott Company, Philadelphia, p.l.

Chancharoensin, S. and Bhumiratana, A. (1983). Production of L-lysine by homoserine auxotrophic mutant of *Corynebacterium glutamicum.* (HOM⁻) *Thai J. Agric. Sci* **16**(4),315

Chatterjee, M., Chatterjee, S.P. and Banerjee, A.K. (1990). Producttion of L-lysine by double auxotrophic and AEC resistant mutants of *Bacillus megaterium Res. Ind.*

35(2), 133.

Coello, N., Pan, J.G. and Lebeault, J.M. (1992). Physiological aspects of L-lysine production: effect of nutritional limitations on a producing strain of *Corynebacterium glutamicm. Appl. Microbiol. Biotechnol.* **38**(2),259.

Cohen, G.N. and Patte, J.C. (1963). Aspects of the regulation of amino acid biosynthesis in a branched pathway. *Cold. Spring Harbor Symp. Quant. Biol.* **28**,513

Cohen, G.N. Patte. J.C. and Truffa-Bachi, P. (1965). Parallel modifications caused by mutations in two enzymes concerned with the biosynthesis of threonine in *Escherichia coli. Biochem, Biophys. Res. Commun.* **19**(4),546.

Cohen, G.N., Stanier, R.Y. and Le Bras, G. (1969). Regulation of the biosynthesis of amino acids of the aspartate family in coliform bacteria and pseudomonads. *J. Bacteriol.* **99** (3),791.

Collins, M.D., and Cummins, C.S. (1986). Genus *Corynebacterium.* In Bergey's Manual of Systematic Bacteriology, Vol. 2. Williams and Wilkins, Baltimore, p. 1266.

Collins, M.D., Goodfellow, M. and Minnikin, D.E.(1982a) A survery of the structure of mycolic acids in *Corynebacterium* and related taxa *J. Gen. Microbiol* **128**, 129.

Collins, M.D., Goodfellow, M. and Minnikin, D.E.(1982b). Fatty acid composition of some mycolic acid containing coryneform bacteria. *J. Gen. Microbiol.* **128**, 2503

Conklin, M.E. (1934). Mercurochrome as a bacteriological stain. *J. Bacteriol* **27**, 30.

Connor, J.M. (2008). Evolution of the global lysine industry – 1960 – 2000, *Purdue Journal Paper* 16283, pp 1 – 28.

Considine, D.M. (1974). Chemical and process technology encyclopdia. McGraw-Hill Book Company, New York, p. 95.

Cremer, J., Eggeling L, Sahm, H. (1991). Control of the lysine biosynthesis sequence in Corynebacterium glutamicum as analyzed by overexpression of the individual corresponding genes. *Appl. Environ. Microbiol.* **57**(6), 1746 – 1752.

Cremer, J., Treptow, C, Eggeling L, Sahm, H. (1988). Regulation of enzymes of lysine biosynthesis in Crynebacterium glutamicum..*J. Gen. Microbiol.* **134**(Pt12), 3221 – 3229.

Dagley, S. and Johnson, A.R. (1956). Appearance of amino acids and peptide in culture filtrates of microorganisms growing in mineral salt medium. *Biochem. Biophys. Acta.* **21**,270

Dagley, S., Dawes, E.A. and Morrison, G.A. (1950). Production of amino acids in synthetic media by *Escherichia coli* and *Aerobacter aerogenes*. *Nature* **165**, 437.

Daoust, D.R. (1976). Microbial synthesis of amino acids. In Industrial Microbiology. Eds. Miller, B.M. and Litsky, W. McGraw-Hill Book Company, New York, p. 106.

Datta, P.K. and Gest, H. (1964). Control of enzyme activity by concerted feedback inhibition. *Proc. Natl. Acad. Sci. U.S.* **52** (4), 1004.

Davis, B.D. (1952). Biosynthetic interrelations of lysine, diaminopimelic acid, and threonine in mutants of *Escherichia coli, Nature*, **169**, 534

Davis B.D. and Mingioli, E.S. (1950). Mutants of *Escherichia coli*, requiring methionine or vitamin B_{12} *J. Bacteriol.* **60**, 17

De Ley, J. (1970). Re-examination of the association between melting point, buoyant density, and chemical base composition of deoxyribonucleic acid. *J. Bacteriol*, **101**, 738.

De Ley, J. and Schell. J. (1963). Deoxyribonucleic acid base composition of acetic acid bacteria. *J. Gen. Microbiol.* **33**, 243.

Demain, A.L. and Masurekar, P. S. (1974). Lysine inhibition of in vivo homocitrate synthesis in *Penicillium chrysogenum. J. Gen. Microbiol.* **82**, 143.

DeMuelenaere, H.J.H., Chen, M.L. and Harper, A.E. (1967). Assessment of factors influencing estimation of lysine availability in cereal products. *J. Agr. Food Chem.* **15**(2), 310.

Dewey, D.L. and Work, E. (1952). Diaminopimelic acid and lysine. *Nature* **169**, 533.

Drechsel, E. (1889). Zur Kenntniss der Spaltungsprodukte des Caseins (Eng: Studies on the degradation products of casein). *J. Prakt. Chem.* **39**, 425.

Drechsel, E. (1890). Ueber die Bildung von Harnstoff aus Eiweiss (Eng: Studies on the formation of urea form protein). *Berichte*, **23**, 3096

Drechsel, E. (1891). Der Abbau der Eiweissstoffe (Eng: Studies on the degradation of proteins) *Arch. Anat. Physiol. Abt.* **248**, 78

Drechsel, E. and Kruger, T.R. (1892), Zur Kenntniss des lysins (Eng: Studies on lysine) *Berichte*, **25** 2454.

Dulaney, E.L., Bilinski, E. and McConnell, W.B. (1956). Extracellular organic nitrogen in *Ustilago manydis* fermentation broths. *Can. J. Biochem. and Physiol.* **34**, 1195.

Dulaney, E.L. (1957). Formation of extracellular lysine by *Ustilago maydis and Gliocladium sp. Can. J. Microbiol* **3**, 467.

Dunce, M., Liepins, G, Mezina, G. and Viesturs, U. (1980). L-lysine. *U.S.S.R. Patent* 675, 980.

Eggeling, L. and Sahm, H. (1999). L-glutamate and L-lysine; traditional products with impetuous development. *Appl. Microbiol. Biotechnol,* **52**, 146 – 153.

Eggeling, L. and Bott, M.(2005). Handbook of Corynebacterium glutamicum. CRS Press LLC, Boca Raton.

Ellinger, A. (1899). Zur Constitution des lysins (Eng: On the structure of lysine). *Berichte,* **32**, 3542.

Ellinger, A. (1900). Die Constitution des Ornithins und des Lysine, zugleich ein Beitrag zur Chemie der Eiweissfaulniss (Eng: Studies on the constitution of ornithine and lysine, together with a report on the chemistry of protein degradation). *Z. Physiol. Chem.* **29**, 334.

Ericson, L.E. and Kurz, W.G. (1962). Microbial production of amino acid. I Synthesis of lysine and threonine by *Ustilago sp.* *Biotech, Bioeng.* **4**, 23.

Fan. C., Chen, L. and Zheng, S. (1988). Characteristics of L-lysine high yielding strain FML 8611. Weishengwuxue Zazhi, **8**(3), 11. Chem. Abs. **110**: 171711b (1989).

Feldberg, C. and Hetzel, C.P. (1958a). How lysine ups protein value of cereal foods. *Food Eng.* **30**, 110.

Feldberg, C. and Hetzel, C.P. (1958b). The role of lysine in cereal food products. *Food Technol.* **10**, 496.

Fisher, P.J. and Conn, J.E. (1942). A flagella staining technique for soil bacteria. *Stain Technol.* **17**, 117.

Fischer, E. and Weigert, F. (1902). Synthese der a,e-Diaminocapronsaure - inactives lysine (Eng: Synthesis of a, e-Diaminocaproic acid - inactive lysine). *Berichte.* **35**, 3772.

Flodin, N.W. (1997). The metabolic roles, pharmacology, and toxicology of lysine. *J. Am. Coll. Nutr,* **16**, 7 – 21.

Flodin, N.W. (1993). Lysine supplementation of cereal food: a retrospective. *J. Am. Coll. Nutr,* **12**, 486 – 500.

Gaillardin, C.M. Poirier, L. Ribet, A.M. and Heslot, H. (1979). General and lysine specific control of Saccharopine dehydrogenase levels in the yeast *Saccharomycopsis lipolytica. Biochimie.* **61**(4), 473

Gilhoe, D.P., Friede, J.D. and Henderson, L.M. (1968). Effect of hydroxylysine on the biosynthesis of lysine in *Streptococcus faecalis   J. Bacteriol.* **95**(3), 856.

Gilvarg, C. (1958). The enzymatic synthesis of diaminopimelic acid. *J Biol. Chem.* **233**, 1501.

Gilvarg, C. (1960), Biosynthesis of diaminopimelic acid. *Federation Proc.* **19**, 948.

Goodfellow, M., Collins, M.D. and Minnikin, D.E. (1976). Thin Layer charomatographic analysis of mycolic acid and other long chain components in whole organism methanolysates of coryneform and related taxa. *J. Gen. Microbiol.* **96**, 351.

Goodfellow, M. and Schaal, K.P. (1979). Identification methods for Nocordia. Actinomadura and *Rhodococcus*. In Identification methods for microbiologists, 2nd ed, Eds, Skinner, F.A. and Lovelock, D.W. the Society for Applied Bacteriology Technical Series No. 14 Academic Press London p. 261.

Griffith, R. S, Norins, A.L. Kagan, C. (1978). A multicentered study of lysine therapy in Herpes simplex infection. *Dermatologica,* **156**, 257 – 267.

Griffith, R. S, Walsh D.E, Myrmel K.H. (1983). Success of L-lysine therapy in frequently recurrent Herpes simplex infection. Treatment and prophylaxis *Dermatologica,* **175**, 183 – 190.

Grivina. P. (1967). Characteristics of the auxotrophic mutants of Brevibacterium strain 22. *Latv. PSR Zinat, Akad. Vestis* **6**, 44 Chem Abs. **67**:79873b(1967).

Gross, M. (1952). Rapid staining of acid fast bacteria. *Am. J. Clin. Pathol.* **22**, 1034.

Guha, A., Mishra, A.K. and Nanda, G. (1982). Strain improvement for L-lysine production. *Trans. Bose. Res. Inst.* **42**(2), 57.

Gunji, Y and Yasueda, H. (2006). Enhancement of L-lysine production in methylotroph *Methylophilus methylotropus* by introducing a mutant LysE exporter. J. Biotechnol. **127**(1), 1 – 13.

Hagino, H. and Nakayama, K. (1970) Accumulaion of a,e-diaminopimelic acid by Lysine auxotrophic mutants of *Brevibacterium ammoniagenes. Nippon Nogei Kagaku Kaishi.* **44**(9), 422 Chem. Abs. **74**:39527e (1971).

Hallaert, J., Morel de Westgaver, C. and Vandamme, E.J. (1987). L-lysine fermentation by aminoethylcysteine resistant *Corynebacterium glutamicum* mutants. *Meded. Fac. Landbouwwet., Riksuniv. Gent.* **52** (4B), 1901.

Hanel, F., Hilliger, M. and Grafe, U. (1981). Effect of oxygen limitation on cellular L-lysine pool and lipid spectrum of *Corynebacterium glutamicum. Biotechnol. Lett* **3**(9),

461.

Hassan, W.H.W. and Rogers, P.L. (1990). Production of L-lysine by fermentation. *Proc. Malays. Biochem. Soc. Conf.* 14th, 176.

Hartmann, M., Tauch, A., Eggeling, L., Bathe, B., Mockel, B, Puhler, A, Kalinowski, J. (2003). Identification and characteriazation of last two unknown genes, dapC and dapF., in the succinylase branch of the L-lysine biosynthesis of Corynebacterium glutamicum. *J. Biotechnol.* **104**, 199 – 211.

Hayashi, M., Mizoguchi, H., Ohnishi, J., Mitsuhashi, S., Yonetani, Y., Hashimoto, S. and Ikeda, M. (2006). A leuC mutant leading to increased L-lysine production and rel independent global expression change in Corynebacterium glutamicum. *Applied Microbiol. Biotechnol.* **72**(4), 783 – 789.

Hedin, S.G. (1895). Eine methode, das lysine, zu isoliren, nebst einigen Bemerkungen uber des Lysatinin (Eng: A method for the isolation of lysine, and some comments of lystainin). *Z. Physiol. Chem.* **21**, 297.

Henderson, Y. (1900), Ein Beitrag zur Kenntniss der Hexonbasen (Eng: Studies on hexonbases) . *Z. Physiol. Chem.* **29**, 320.

Hermann, M., Thevenet, N.J., Coudert-Maratier, M.M. and Vandecasteele, J.P. (1972). Concequences of lysine oversynthesis in Pseudorunal mutants insensitive to feedback inhibition. Lysine exceretion or endogenous induction of a lysine-catabolic pathway. *Eur. J. Biochem.***30**(1), 100.

Hermann, T. (2003). Industrial production of amino acids by coryneform bacteria. *J. Biotechnol.* **104**, 155 – 172.

Hilliger, M., Fuchs, R., Kreibich, G., Menz, J., Weise, H., Harzfeld, G. and Boettcher, J. (1990) Lysine fermentation using molasses as carbon source. *Gen(East) patent, DD.* **281**, 451 *Chem Abs.* **114**:141621 n (1991).

Hillger, M., Gross, H.H., Menz, H.J., Bormann, E.J., Grosse, P., Miosga, N., Kreibich, G. and Fuchs, R. (1991), Optimization of lysine manufacture by fermentaion. *Ger(East). Patent, DD* 290,213. Chem. Abs. **115**:112808 u (1991).

Hilliger, M., Kreibich, G., Roth, M., Fuchs, R., Menz. J., Rau, A., Grosse., H.H., Schiller, E. and Seeger, B. (1992). Mauafacture of lysine with *Corynebacterium glutaicum*. *Ger(East). Patent. DD*. 300,853. Chem Abs. **118**:5744k (1993).

Hilliger, M., Prauser, H., Menz, H.J. and Grosse, H.H. (1989b). Manafucture of lysine with *Oerskovia*. *Ger(East). Patent. DD*. 273,647. Chem. Abs. **113**:4645p (1990).

Hilliger, M., Kreibich, G., Thiele, G., Boettger, D., Bergter, F., Jagusch, L. and Schoenherr, W (1989 a). Fermentation of lysine. *Ger(East). Patent* DD 268,834. Chem. Abs. **112**:117387 d (1990).

Hirao, T., Nakano, T., Azuma, T., Sugimoto, M. and Nakanishi, T. (1989). L-lysine production in continuous culture of an L-lysine hyperproducing mutant of *Corynebacterium glutamicum. Appl. Microbiol. Biotechnol.* **32**(3), 269.

Hirose, Y. and Shibai, H. (1980). Amino acid fermentation, *Biotechnol. Bioeng.* **22**(Suppl.l), 111.

Hirose, Y. and Shibai, H. (1985). L-Glutamic acid fermentation. In: Moo-Young M, editor, Comprehensive biotechnology, 3, 595 – 600.

Hiss, P.J., Jr. (1905). A contribution to the physiological differentiation of *Pneumococcus* and *Streptococcus*, and to methods of staining capsule. *J. Expt. Med.* **6**, 317.

Hitchcock. M.J. M. and Hodgson, B. (1976). Lysine and lysine plus threonine inhibitable aspartokinases in *Bacillus brevis, Biochim. Biophys. Acta,* **445**(2), 350.

Hitchcock, M.J.M. Hadgson. B. and Linforth, J.L. (1980). Regulation of lysine and lysine plus threonine inhibitable aspartokinases in *Bacillus brevis. J. bacteriol* **142**(2), 424.

Hlasiwetz, H. and Habermann. J. (1873). Ueber die Proteinstoffe (Eng: Studies on the proteinous materials). *Ann.* **169**, 150

Hogg, R.W. and Broquist, H.P. (1968). Homocitrate formation in Neurospora crassa. *J. Biol.Chem.* **243**(8), 1839.

Hucker, G.J. and Conn. H.J. (1923). Methods of Gram staining, *N.Y.State Agr. Expt. Sta. Tech. Bull.* 129.

Hucker, G.J. and Conn. H.J. (1927). Further studies on the methods of Gram staining *N.Y. State Agr. Expt. Sta. Tech. Bull.* 128.

Hundley, J.M. and Ing, R.B. (1956). Algae as sources of lysine and threonine in supplementing wheat and bread diets. *Science.* **124**, 536

Huang, H.T. Griffin, J.M. and Fried, J.H. (1960). Fermentation process for the production of diaminopimelic acid, *U.S. patent.* 2,955,986.

Ikeda, M. (2003). Amino acid production. In: Faurie R, Thommel J (eds), *Adv. Biochem. Eng. Biotechnol.* Vol **79**. Microbial production of L-amino acids. Springer, Berlin Heidelberg, New York, pp 1 -35

Ikeda, M., Ohnishi, J. and Hayashi, M. . (2006). A genome-based approach to create a minimally mutated *Corynebacterium glutamicum* strain for efficient L-lysine production, *J. Ind. Microbiol. Biotechnol.* **33.** 610 - 615

Inuzuka, K. and Hamada, S. (1976). Process for the production of L-lysine *U.S. Patent.* 3,959,075.

Ishikawa, K., Asahara, T., Gunji, Y., Yasueda, H. and Asano, K. (2008a). Disruption of metF increased L-lysine production by *Methylophilus methylotrophus* from methanol. Biosci. Biotechnol Biochem. **72**(5), 1317 – 1324.

Ishikawa, K., Asahara, T., Gunji, Y., Yasueda, H. and Asano, K. (2008b). Improvement of L-lysine production by *Methylophilus methylotrophus* from methanol via the Entner-Doudoroff pathway, originating in *Escherichia coli.* Biosci. Biotechnol Biochem. **72**(10), 2535 – 2542..

Jensen, H.J. (1952). The coryneform bacteria. *Ann Rew. Microbiol* **6**, 77.

Kagan, C. (1974). Lysine therapy for herpes simplex. *The Lancet, January* 26, 137.

Kalinowaski J., Bachmann B., Thierbach G. and Puhler A. (1990). Aspartokinase gene lysC alpha and lysC beta overlap and are adjacent to the aspartate beta-semialdehyde dehydrogenase gene asd in *Corynebacterium glutamicum. Mol. Gen. Genet.,* **224**(3), 317 – 324.

Kalinowaski J., Bathe, B., Bartels D., Bischoff N., Bott, M., Burkovski, A., Dusch, N., Eggeling, L., Eikmanns, B., Gaigalat, L., Goesmann, A., Hartmann, M., Huthmacher, K., Kramer, R., Linke, B., McHardy, A.C., Meyer, F., Mockel, B., Pfefferle, W., Puhler A, Rey D.A., Ruckert, C., Rupp, O., Sahm, H., Wendisch, V.F, Wiegrabe, I. and Tauch, A. (2003). The complete Corynebacterium glutamicum ATCC 13032 genome sequence and its impact on the production of L-aspartate amino acids and vitamins. *J. Biotechnol.* **104**, 5 – 25.

Kaneko, T. and Co-Workers. (1974). Synthetic production and utilization of amino acids. Kodansha, Tokyo, Japan, through Kirkothmer Encyclopedia of Chemical Technology, John Wiley and Sons, New York, 3rd ed, vol. 2, p. 376 (1978).

Karlstrom.O. (1965). Methods for the production of mutants suitable as amino acids fermentation organisms. *Biotechnol. Bioeng.* **7**, 245.

Kase, H. and Nakayama, K. (1970). Production of a,e-diaminopimelic acid by lysine auxotrophs of various bacteria. *Nippon Nogei Kagaku Kaishi.* **44**(10) 457, Chem. Abs. **74**:39601z (1971).

Kawahara, Y., Yoshihara, Y., Ikeda. S., Yoshii, H. and Hirose, Y. (1990). Stimulatory effect of glycine betaine on L-lysine fermentation. *Appl. Microbiol Biotechnol* **34**(1), 87.

Keddie, R.M. and Bousfield, I.J. (1980). Cell wall composition in the classification and identification of coryneform bacteria. In Microbiological Claasification and Identification. Eds. Goodfellow, M. and Board. R.G. Acadermic Press London, p. 167.

Keddie, R.M. and Cure. G.L. (1977). The cell wall composition and distribution of free mycolic acids in named strains of coryneform bacteria and in isloates from various natural source. *J. Appl. Bacteriol.* **42**, 229

Keddie, R.M. and Cure. G.L. (1978). Cell wall composition of coryneform bacteria. In coryneform bacteria. Eds. Bousfield I.J. and Callely. A.G. Academic Press, London. p. 47.

Keleti., G. and Lederer, W.H. (1974). Handbook of Micromethods for the Biological Sciences. Van Nostrand Reinhold Company, New York. p. 10.

Kim, B.H. Seong, B.L. Mheen, T.I. and Han, M.H. (1981) Studies on microbial penicillin amidase. I. Optimization of the enzyme production from *Escherichia coli. Korean J. Appl. Microbiol. Bioeng.* **9**(1), 29.

Kinoshita, S. (1959). The production of amino acids by fermentation processes. In advances in Applied Microbiology. Ed. Umbreit, W.W. vol. 1 Academic Press, New York, p. 201.

Kinoshita, S. (1985). Glutamic acid bacteria. In Biology of Industrial Microorganisms. Eds. Demain, A.L. and Solomon, N.A. The Benjamin/Cummings Publishing Company. Inc. California, p. 115.

Kinoshita, S., Itagaki, S. and Nakayama, K. (1960). Amino acids **2**, 42, through *J. Gen, Appl. Microbiol.* **13**, 279 (1967)

Kinoshita, S. and Nakayama, K. (1978). Amino acids. In Economic Microbiology, vol. 2, Primary products of metabolism. Ed. Rose, A.H. Academic Press, London, P. 209.

Kinoshita, S., Nakayama, K. and Akita, S. (1958 c). Taxonomical study of glutamic acid accumulating bacteria, *Micrococcus glutamicus* nov. sp. *Bull. Agr. Chem. Soc. Japan.* **22**(3), 176.

Kinoshita, S., Nakayama, K, and Kitada, S. (1958 b). L-lysine production using microbial auxotroph. *J. Gen. Appl. Microbiol.* **4**(2), 128.

Kinoshita, S., Tanaka, K, Udaka, S. Akita, S., Saito, T. and Iwazaki, T. (1958 a). Glutamic acid fermentation. *Hakko Kyokaishi* **16**, 1.

Kinoshita, S., Nakayama, K. and Kitada, S. (1961). L-lysine manufacture by fermentation. Japan Patent, 6499. Chem. Abs. **56**:1844i (1962).

Kinoshita, S. Tanaka, T., Udaka, S. and Akita, S. (1957 b). Glutamic acid fermentation. *Proc. Symp. Enzyme Chem.* **2**, 464

Kinoshita, S., Udaka S. and Shimono, M. (1957 a). Studies on the amino acid fermentation Part 1. Production of L-glutamic acid by various microoogranisms, *J. Gen. Appl Microbiol.* **3**(3), 193.

Kircher, M. and Pfefferle, W. (2001). The fermentative production of L-lysine as an animal feed additive. *Chemosphere.* **43**, 27 – 31.

Kita, D.A. (1957). Production of glutamic acid by *Cephalospotium sp. U.S. Patent* 2,789,939.

Kita, D.A. and Huang, H.T. (1958). Fermentation process for the production of L-lyine. *U.S. Patent* 2,841, 532.

Kleemann, A., Leuchtenberger, W., Hoppe, B. and Tanner, H, (1985). Amino acids, In Ulmann's Encycolpedia of Industrial Chemistry, 5th ed. vol. A2 Ed. Gerhartz, W. VCH Verlagsgesellschaft mbH. Germany, p.57.

Koike, M. and Reed, L.J. (1960). a-keto acid dehydrogenation complex. II. The role of protein bound lipoic acid and flavin adenine dinucleotide. *J. Biol. Chem.* **235**, 1931.

Komagata, K. and Suzuki. K. (1987). Lipid and cell wall analysis in bacterial systematics. *Methods in Microbiol.* **19**, 161.

Komagata, K., Yamada, K. and Ogawa, H. (1969). Taxonomic studies on coryneform bacteria. I. Division of bacterial cells. *J. Gen. Appl. Microbiol.* **15**, 243.

Komatsu, Y. and Kaneko, T. (1980). Deoxyribonucleic acid relatedness between some glutamic acid producing bacteria. Report of the Fermentation Research Institute. Tsukuba, Japan, **55**, 1.

Komatsubara, S., Kisumi, M. and Chibata, l. (1979). Participation of lysine sensitive aspartokinase in threonine production by S-2-amino-ethy cysteine resistant mutants of *Serratia marcescens. Appl. Environ Microbiol.* **38**(5), 777.

Komatsubara, S., Kisumi, M., Murata, K. and Chibata, I. (1978). Threonine production by regulatory mutants of Serratia marcesscens. *Appl. Environ. Microbio.* **35**(5), 834.

Komura, I., Yamada, K. and Otsuka, S. (1975). Taxonomic significance of phospholipids in coryneform and nocardioform bacteria, *J. Gen. Appl. Microbiol.* **21**, 251.

Kurihara, S., Shi, H., Ken, Y., Araki, K., Shi, M. and Akeyama, K. (1972). Process for producing L-lysine by fermentation. *U.S. Patent.* 3,687,810.

Kurtz, M. and Bhattacharjee, J.K. (1975). Biosynthesis of lysine in *Rhodotorula glutinis*: role of pipecolic acid. *J. Gen. Microbiol.* **86**, 103.

Kurz, W.G. and Ericson. L.E. (1962), Microbial production of amino acids. II. The influence of carbon and nitrogen sources and metal ions on the growth of *Usteago maydis* on lysine and threonine production, *Biotechnol. Bioeng.* **4**, 37.

Laskin, A.T. (2008): In – Adevances in applied microbiology, published by the Academic Press, London, UK, **63**, pp.238.

Lederberg, J. and Leaderberg, E.M. (1952). Replica plating and indirect selection of becterial mutants. *J. Bacteriol.* **63**, 399.

Lee, I.S. and Cho, J. (1994). Continuous fermentation of L-lysine by immobilized *Corynebacterium glutamicum*, Han'guk *Yangyang Sikyong Hakhoechi* **23**(2), 322. Chem Abs. **121**:228937 p (1994).

Leuchtenberger, W. (1996). Amino acids: technical production and use. In: Roehr M (eds) *Biotechnology,* 2nd edition, vol. 6. Products of primary metabolism. VCH, Weinheim, pp 465 – 502.

Lin, D., Pan, J., Li. B., Chen. T., Wang, Y. and Yu, R. (1990). Studies on the selection and fermentation conditions of L-lysine producing strain. *Gongye Weishengwu* **20**(4), 16, Chem Abs. **114**:4809 r (1991).

Liu, Y.T. (1987). Studies on the fermentative production of L-lysine. 6. An improvement of L-lysine producing strain by mutation of regulatory gene. *Taiwan Tang Yen Chiu So Yen Chiu Hui Pao.* **116**, 39 Chem. Abs, **109**:53147 d (1988).

MacFaddin, J.F. (1976). Biochemical Tests for Identification of Medical Bacteria, The Williams and Wilkins Company. Baltimore. P.1.

Mahmood, Z.A., Shaikh, D., Shaikh, M.R., Naqvi, B.S. and Zoha, S.M.S. (1996). Production of L-lysine by auxotrophi mutants. *Acta Manilana.* **44**, 31 – 36.

Malumbres, M. and Martin, J.F. (1996). Molecular control mechanisms of lysine and threonine biosynthesis in aminoacid-producing *Corynebacterium*: Redirecting carbon flow. *FEMS Microbiol. Lett.* **143**, 103 – 114.

Maragoudakis, M.E., Holmes, H. and Strassman. M. (1967). Control of lysine biosynthesis in yeast by a feedback mechanism. J. Bacteriol. **93**(5), 1677.

Maria, C.F. and Duarte J.C. (1992). Amino acid accumulation by an analog sensitive mutant of *Corynebacterium glutamicum. Biotechnol. Lett.* **14**(11),1025

Marmur, J. and Doty, P. (1962). Determination of the base composition of deoxyribonucleic acid from its thermal denaturation temperature. *J. Mol. Biol.* **5**,109.

Masurekar, P.S. and Demain, A.L. (1974). Insensitivity of homocitrate synthase in extracts of *Penicillium chrysogenum* to feedback inhibition by lysine. *Appl. Microbiol.* **28**(2), 265.

McBeath, M. and Pauling, L. (1993). A case history: lysine / ascorbate-related amelioration of angina pectoris. *J. Orthomolecular Med.* **8**, 77 -78.

Miller, B. M. and Litsky, W. (1976). Industrial Microbiology. McGraw-Hill Book Company, New York. p. xi.

Minnikin, D.E., Alshamaony, L. and Goodfellow, M. (1975). Differentiation of *Mycobacterium, Nocardia*, and related taxa by thin layer chromatographic analysis of whole organism methanolysates. *J. Gen. Microbiol.* **88**, 200.

Minnikin, D.E., Goodfellow, M. and Collins, M.D.(1978). Lipid composition in the calssification and identification of coryneform and related taxa. In Coryneform bacteria, Eds. Bousfield, I.J. and Callely, A.G. Academic Press, Inc. London. p. 85.

Misra, A.K., Dasgupta, J., Malaviya, A. and Vora, V.C. (1979). Fermentative production of L-lysine. *Indian J. Microbiol.* **19**(3), 130.

Misra, A.K., Dasgupta, J. and Vora, V.C. (1980). Microbial production of L-lysine. *J. Chem. Tech. Biotech.* **30**, 453.

Mitchell, H.K. and Houlahan, M.B. (1948). An intermediate in the biosynthesis of lysine in *Neurospora. J. Biol. Chem.* **174**, 883

Moat, A.G. (1979). Microbiol Physiology, Jhon Wiley and Sons, New York, p. 161.

Morris, D.L. (1948). The quantitative determination of carbohydrates with Dreywood's anthrone reagent. *Science.* **107**, 254.

Morton, A.G. and Broadbent, D. (1955). The formation of extracellular nitrogen compounds by fungi, *J. Gen. Microbiol.* **12**, 248.

Moss, J. and Lane, M.D. (1971). The biotin dependent enzymes. In Advances in Enzymology. Ed. Miester. A., Jhon Wiley and Sons Inc. New York, **35**, 321.

Muradyan, A.G., Dzhilavyan, L.R., Dzhamgaryan, S.M., Madatovyan, A.M. and Oganesyan, M.G. (1980). Study of the sorption of L-lysine and some inorganic cations on sulfonated cation exchanger KU-2XB. *Arm. Khim. Zh* **33**(6), 473. Chem. Abs, **93**:166030 e (1980).

Murakami, Y., Miwa, H. and Nakamori, S. (1991) Manufacture of lysine with analog resistant *Corynebacterium Fr. Demande FR* 2,661,191.

Nakayama, K. (1972 a). Lysine and diaminopimelic acid. In The Microbial Prodiction of
Amino Acids. Eds. Yamada, K., Kinoshita, S., Tsunoda, T. and Aida, K.,
Kodansha Ltd, Tokyo. P. 369.

Nakayama,K. (1972 b). Microorganisms in amino acid fermentation. *Proc. IV IFS:
Ferment. Technol. Today*, P. 433.

Nakayama, K. (1976). The production of amino acids. *Process Biochem.* **3**, 4

Nakayama, K. (1983). Amino acids. In Prescott and Dunn's Industrial Microbiology. 4th ed.
Ed. Reed. C. AVI Publishing Company, Inc, Connecticut. p.748.

Nakayama, K.(1985). Lysine. In Comprehensive Biotechnology. Ed. Moo-Young, M. Vol.
3. Pergamon Press, Oxford,. pp. 607 - 620.

Nakayama, K. and Araki, K. (1973). Process for producing L-lysine *U.S. Patent*
3,708,395

Nakayama, K. and Araki, A. (1981). Process for producing L-lysine by fermentation.
Japan Patent. 818692, through Comprehensive Biotechnology vol. 3 Ed. Moo-
Young, M. Pergamon Press Oxford.

Nakayama, K., Araki, K. and Tanaka, Y. (1979). Process for the production of L-lysine by
fermentation *U.S. Patent.* 4,169,763.

Nakayama, K. and Kinoshita, S. (1961 a). Studies on lysine fermentation. II α, e
Diaminopimelic acid and its decarboxylase in lysine producing strain and parent
strain. *J. Gen. Appl. Microbiol.* **7**, (3), 155.

Nakayama, K. and Kinsohita, S. (1961 b). Studies on lysine fermentation III. a,e
Diaminopimelic acid accumulation and diaminopimelic acid decarboxylase. *J. Gen.
Appl. Microbiol.* **7**(3), 161.

Nakayama, K. and Kinoshita, S. (1961 c). Lysine fermentation IV. Culture conditions for
lysine accumulation by a homoserine free mutant of *Bacillus subtilis. Nippon
Nogeikagaku Kaishi.* **35**, 119. Chem. Abs. **59**:15617 h (1963).

Nayakama, K., Kitada, S. and Kinoshita, S. (1961 a) Studies on lysine fermentation. I. The
control mechanism on lysine accumulation by homoserine and threonine, *J. Gen.
Appl. Microbiol.* **7**(3), 145.

Nakayama, K., Kitada, S., Sato. Z. and Kinoshita, S. (1961 b). Induction of nutritional
mutants of glutamic acid bacteria and their amino acid accumulation, *J. Gen.
Appl. Microbiol.* **7**(1), 41.

Nakayama, K., Tanaka, H., Hagino, H. and Kinoshita, S. (1966). Studies on lysine fermentation. Part. V. Concerted feedback inhibition of aspartokinase and the absence of lysine inhibition on aspartic semiladehyde pyruvate condensation in *Micrococcus glutamicus, Agr. Biol. Chem.* **30**(6), 611.

Nasri, M., Dhouib, A., Zorguani, F., Kriaa, H. and Ellouz, R. (1989). Production of lysine by using immobilized living *Corynebacterium sp* cells. *Biotechnol. Lett.* **11**(2), 865.

Neish, A.C. (1952), Analytical methods for bacterial fermentaiton. Manual, published by the National Research Council of Canada. P.l.

Nelson, G.E.N., Anderson, R.F. Rhodes, R.A., Shekleton, M.C. and Hall H.H. (1960). Lysine, methionine, and tryptophan content of microorganisms, II, *Yeasts, Appl. Microbiol.* **8**, 179

Nhan, H.B., Siehr, D.J. and Findley, M.E. (1976). Studies on the rate of lysine production by *Brevibacterium lactofermentum* from glucose. *J. Gen. Appl. Microbiol.* **22**,65.

Nishiyama, M. and Nishida, H. (2000). What is characteristic of fungal lysine synthesis through the α - aminoadipate pathway.

Nomura, Y., Umeki, H., Fukuoka, S., Iwahara, M. and Hongo, M. (1980). Studies on L lysine fermentation. On various properties of *Bravibacterium flavum* QL-5. *Kenkyu Hokou - Kumamoto Kogyo Daigaku,* **5**(1), 183 Chem. Abs. **93**:130486 u (1980).

n.d. (1953). Difco Manual of Dehydrated and Culture Media and Reagents for microbiological and clinical laboratory procedures. 9th ed. Difco Laboratories Inc. Detroit , Michigugan. p. 233.

n.d. (1996). Drug, facts and comparisons. Oral nutritional supplements-amino acids, Loose - leaf drug information service. Ed. Oline, B.R. Facts and comparisions, Inc. St. Louis , 53b.

n.d. (2007). L-lysine – Health Care Industry, *Alternative Medicine review.* pp 1 – 10.

n.d. (2009). Foreign Trade Statistics of Pakistan. Imports. Published by Economic affairs and statistics division Government of Pakistan.

n.d. (2010). Physician's Desk Reference. Product information – L-lysine, 64, pp 3455.

Obeta, J.A.N. and Ekwealor, I.A. (2006). Screening of UV-irradiated and S-2-aminoethyl-L cysteine mutants of Bacillus megaterium for improved lysine accumulation (2006).

African J. Biotechnol. **5**(22), 2312 – 2314.

Osborne, T.B. and Mendel, L.B. (1914). Amino acids in nutrition and growth *J. Biol. Chem.* **17**, 325.

Osborne, T.B. and Mendel, L.B. (1919). The nutritive value of the wheat kernel and its milling products. *J.Biol Chem.* **37**, 557.

Patek, M., Bilic, M., Krumbach, K., Eikmanns, B., Sahm, H. and Eggeling, L. (1997). Identification and transcriptional analysis of the dapB-orf2-dapA-orf4 operon of Corynebacterium glumtamicum, encoding two enzymes involved in L-lysine sysnthesis. *Biotecnol. Letters.* **19**(11), 1113 – 1117.

Pauling, L. (1991). Case report: lysine / ascorbate-related amelioration of angina pectoris. *J. Orthomolecular Med.* **16**, 144 – 146.

Pauling, L. (1993). Third case report on lysine- ascorbate-related amelioration of angina pectoris. *J. Orthomolecular Med.* **8**, 137 – 138.

Paulus, H. and Gray, E. (1964), Multivalent feedback inhibition of aspartokinase in *Bacillus polymyxa. J. Biol. Chem.* **239**(11), 4008.

Pavia, D.L., Lampman, G.M. and Kriz Jr., S.G.(1970). Introduction to spectroscopy. Sounders collage publishing company. Winston. P.68.

Pelechova, J., Seifert, R. and Smekal, F. (1983). Biosynthesis of L-lysine from paper hydrolyzate with *Corynebacterium glutamicum* and *Brevibacterum sp. Kvasny Prum* **29**(12), 279. Chem. Abs. **100**:173085 t (1984.

Petrov, D.F., Orobinskii, I.I., Shilova, M.A., Grableva, T.I., Tarakanova, E.G. and Zakirova, T.F. (1965). New Microbiological method of preparation of lysine. *Selektsiya Mikrobov, Akad. Mauk SSSR, Sibrisk. Otd.,* Lab. *Tsitol. Rast. i Apomiksisa Biol. Inst.* 128. Chem. Abs. **64**:14920 c (1966).

Petrov, D.F., Shilova, M.A. and Orobinskii, 1.1.(1963). Preparation of lysine for animals. *U.S.S.R. Patent.* 161, 221. Chem. Abs. **61**:1184 a (1964)

Pfefferle, W., Lotter, H., Friedrich, H, and Degener, W., (1993). Improved yields in amino acid fermentation with coryneiform bacteria. *Ger. Offen. DE.* 4,130,867. Chem.Abs. **118**:190117 u (1993).

Pham,C.B., Mercado, C.J., Matsumura, M. and Kataoka, H.(1990). Continuous lysine fermentation with free cells in stirred tank reactor. *Microb. Util. Renewable Resour.* **7**, 538.

Plachy, J. (1970). Effect of medium composition on the production of lysine by

Corynebacterium sp. 9366-H 454. *Folia Microbiol.* **15**, 347.

Plachy, J., Culik, C., Ulbert, S., Bucko, M., Chromik, J., Miroslav. J., Hofbauer, H., Miklas, E. and Severa, E. (1977). Lysine by fermentation. *Ger (East) Patent.* 140,056. Chem. Abs. **93**:112307g (1980).

Plachy, J. and Ulbert, S. (1987). Application of mutants sensitive to amino acids for preparation of lysine by fermentation. *Kvasny Prum.* **33**(7), 203, Chem. Abs. **109**:71941 n (1988).

Plachy, J. and Ulbert, S. (1988). Using a mutant *Corynebacterium glutamicum* 9366 AEC/100 for preparation of lysine by fermentation. *Kvasny Prum.* **34**(8-9), 239, Chem. Abs. **110**: 73801 m (1989).

Plachy, J., Ulbert, S. and Smekal, F. (1988). An application of auxotrophic-regulartory mutants of *Corynebacterium glutamicum* for preparation of lysine by fermentaion. *Kvasny Prum.* **34**(11), 328. Chem. Abs. **110**: 210858 b (1989).

Pontecorvo, G. (1949). Auxanographic techniques in biochemical genetics *J. Gen. Microbiol.* **3**(1), 122.

Reusser, F., Spencer, J.F.T. and Sallans, H.R. (1957). Essential amino acids in microorganisms. *Can. J. Microbiol.* **3**, 721

Rhodes, R.A., Hall, H.H., Anderson, R.F., Nelson, G.E.N., Shekleten, M.C. and Jackson R.W.(1961). Lysine, methionine, and tryptophan content of microorganisms. III. Molds. *Appl. Microbiol.* **9**, 181.

Rhuland, L.E., Work , E., Denman, R.F. and Hoare, D.S.(1955). The bahavior of the isomers of a,e-diaminopimelic acid on paper chromatography. *J. Am. Chem. Soc.* **77**, 4844.

Richards, M. and Haskins, R. H. (1957). Extracellular lysine production by various fungi, *Can, J. Microbiol.* **3**, 543.

Rodionov, D.A., Vitreschak, A.G., Mironov, A.A. and Gelfand, M.S. (2003). Regulation of lysine biosynthesis and transport genes in bacteria: yet another RNA riboswitch. *Nucleic Acid Research.* **31**(23). 6748 – 6757.

Rodwell, V.W.(1993). Structures and functions of proteins and enzymes. Amino acids. In Harper's Biochemistry. Ed. Robert K. Murray, 23rd ed. Appleton and Lange, Connecticut, p. 23.

Rosner, A. (1975). Control of lysine biosynthesis in *Bacillus subtilis.* Inhibition of diaminopimelate decarboxylase by lysine *J. Bacteriol.* **121**(1), 20.

Rutkov, A. (1983 a). Microbial production of Lysine. Part I. Substrate specificy for growth and lysine productivity of *Croynebacterium glutamicum* ATCC 13286. *Acta Microbiol. Bulg.* **13**, 33 Chem. Abs. **100**: 4730 u (1984).

Rutkov, A. (1983 b). Microbial production of L-lysine. Part II. Inhibitory effect of L threonine and L-lysine on the lysine productivity of *Corynebacterium glutamicum* ATCC 13286. *Acta Microbiol. Bulg.* **13**, 40. Chem. Abs. **100** : 4731v (1984).

Samanta, T.K. and Bhattacharyya, R. (1991). L-lysine production by S-2-aminoethyl-L cysteine-resistant mutants of *Arthrobacter globifermis*. *Folia Microbiol* **36**(1), 59.

Samanta, T.K., Das., Y., Mondal, S. and Chatterjee, S.P. (1988), L-lysine production by auxotrophic mutants of *Arthrobacter globifermis*. *Acta Biotechnol.* **8**(6). 527.

Samejima, H. (1972). Methods for extraction and purification. In The Microbial Production of Amino Acids. Ed. K. Yamada, Kodansha Ltd. Japan, p. 227.

Sano, K. and Shiio, I. (1967). Microbial production of L-lysine.I. Production by auxotrophs of *Brevibacterium flavum*. *J. Gen. Appl. Microbiol.* **13**, 349.

Sano, K. and Shiio, I. (1970). Microbial production of L-lysine. III Production by mutants resistant to S-(2-aminoethyl)-L-cysteine. *J. Gen. Appl. Microbiol.* **16**, 373.

Sassi, A.H., Coello, N., Deschamps, A.M. and Lebeault, J.M. (1990) Effect of medium composition on L-lysine production by variant strain of *Corynebacterium glutamicum* ATCC-21513. *Biotechnol. Lett.* **12**(4), 295.

Sato , Y., Kainuma, M., Nara, S., Terasawa, M. and Yugawa, H. (1990) Manufacture of L lysine with coryneform bacteria. *Japanese Patent.* 0242,995 Chem. Abs. **113**: 22226g (1990).

Schendel, F.J., Bremmon, C.E., Flickinger, M.C., Guettler, M. and Hanson, R.S. (1990). L-lysine production at 50°C by mutants of a newly isolated and characterized methylotrophic *Bacillus sp. App. Environ. Microbiol* **56**(4), 963.

Schrumpf, B. (1991). Lysine formation with *Corynebacterium glutamicum*: analysis of the metabolic flow by determining the cell-internal amino acid concentrations and enzymatic studies. *Forschungszent. Juelich: Bar. Juel* **2478**, 108. Chem. Abs. **116**:37655g (1992).

Schrumpf. B., Eggeling, L. and Sahm, H. (1992). Isolation and prominent characteristics of an L-lysine hyperproducing strain of *Corynebacterium gluamicum*. *Appl. Microboil. Biotechnol,* **37**, 566.

Schrumpf. B., Schwarzer, A, Kalinowaki, J, Puhler, A, Eggeling, L. and Sahm, H. (1991). A functionally split pathway for lysine synthesis in *Corynabacterium glutamicum J. Bacteriol.*, **173**, 4510 – 4516.

Schützenberger. P., (1879). Memoire sur les matieres albuminoides. (Eng: A report on the albuminiod materials). *Ann. Chim. Phys.* **16**(5), 289.

Scrimshow, N.S. and Altschul, A.M. (1971). Amino acid fortification of protein foods, MIT Press, Cambridge, Mass, through Kirkothmer Encyclopedia of Chemical Technology. John Wiley and Sons, New York, 3rd ed. Vol. 2. p. 376 (1978).

Sen, S.K. (1991). Isolation method for lysine-excreting mutants of *Arthrobacter globiformis*. *Folia Microbiol.* **36**(1), 67.

Seto, K. and Harada, T. (1969). Formation of L-lysine form acetic acid by a homoserine auxotroph of *Corynebacterium acetophilum* A-51. *Hakko Kogaku Zasshi.* **47**(9), 558.

Shallenberger, R.S., Acree, T.E. and Lee, C.Y. (1969). Sweet taste of D and L-sugars and amino acids and the steric nature of their chemo-receptor site. *Nature.* **221**, 555

Shaw, J.F. and Smith, W.G. (1977). Studies on the Kinetic mechanism of lysine-sensitive aspartokinase. *J. Biol. Chem.* **252**(15), 5304.

Shiio, I. and Sano, K. (1969). Microbial production of L-lysine. II. Production by mutants sensitive to threonine or methionine, *J. Gen. Appl. Microbiol.* **15**, 267.

Shockman, G.D. (1963). Amino acids. In Analytical Microbiology. Ed. Frederick Kavanage, Academic Press, New York, P. 567.

Siegfried, M. (1891) Zur Kenntniss der Spaltungsproducte. (Eng: Studies on the split products of protein moities). *Berichte.* **24**, 418.

Smekal, F. (1983). Fermentation production of L-lysine. *Czech. CS.* 203, 769. Chem. Abs. **100**:4795u (1984).

Smekal, F., Pelechova, J., Sirochova, E., Zdanova, N. and Leonova, T. (1988). Biochemical and production properties of *Corynebacterium glutamicum* strains. *Kvasny Prum.* **34**(1), 12. Chem. Abs. **109**:91169v (1988).

Soda, K., Takonouchi, E. and Tanaka, H. (1981). Lysine production by yeasts and its mechanism. *Hakko Kogaku Kaishi.* **59**(1), 59. Chem. Abs. **94**:101259 n (1981).

Stakman, E. C. (1964). Applied Microbiology in the future of mankind. In Global Impacts of Applied Microbiology. Ed. Mortimer P. Starr, John Wiley and Sons, Inc., New York. p. 9.

Stanbury, P.F. and Whitaker, A. (1984). Principles of fermetnation technology. Pergamon

Press. Oxford. p. 26.

Stikans, A., Sharifullin, V.N., Emelyanov, V.M. and Valdimirova, I.S. (1991). Intensification of lysine biosynthesis using surfactants. *Izv. Vyssh. Uchebn. Zaved., Khim Khim. Tekhnol.* **34**(6), 108. Chem. Abs. **115**:230438W (1991).

Sukharevich, V.I. Kislukhina, O.V., Tuturina, V.A., Listskaya, T.B. and Medvedeva, N.G. (1992). The influence of systhetic carbohydrates on lysine biosynthesis. *Biotekhnologiya.* **2**, 30. Chem. Abs. **117**:149381s (1992).

Sur, B., Misra, A.K. and Basu, S.K. (1991). Use of mutants deficient to amino acids for production of lysine by fermentation. *Indian Biologist.* **23**(1), 10.

Suzuki, K., Kaneko, T. and Komagata, K. (1981). Deoxyribonucleic acid homologies among coryneform bacteria. *Int. J. System. Bacteriol.* **31**(2), 131.

Suzuki, K. and Komagata, K. (1983). Taxonomic significance of cellular fatty acid composition in some coryneform bacteria. *Int. J. System., Baceriol.* **33**(2), 188.

Takenouchi, E., Tanaka, H. and Soda, K. (1981). Homocitrate synthase of S-(ß aminoethyl)-L-cysteine resistant mutant of *Candida pelliculosa. J. Ferment. Technol.* **59**(6), 429.

Tan, Y., Zhang, X. and Tao, W. (1983). Mechanism of the breading of L-lysine producing bacteria. Breading of drug-resistant *Brevibacterium flavum* starins. *Shipin Kexue,* **41**(1-8), 38. Chem. Abs. **100**:173090 r (1984).

Tanaka, H., Hagino, H. and Nakayama K. (1967). Studies on lysine fermentation . VI. Diaminopimelic acid accumulation by a lysine auxotroph of *Bacillus subtilis. Nippon Nogei Kegaku Kaishi.* **43**(3), 106. Chem. Abs. **67**:8858q(1967).

Tateno, T., Fukuda, H. and Kondo A. (2007). Production of L-lysine from starch by Corynebacterium glutamicum displaying   - amylase on its cell surface, *Appl. Microbiol. Biotechnol.,* **74**, 1213 – 1220.

Tatum, E.L. (1945). X-ray induced mutant strains of *Escherichia coli. Proc. Natl. Acad, Sci. U.S.* **31**, 215.

Tauro, P., Ramachandra Rao, T.N., Johar, D.S., Sreenivasan. A. and Subrahmanyam, V. (1963). L-lysine production by *Ustilaginales fungi. Agr. Biol. Chem.* **27**, 227.

The Pharmacopoeia of Japan. (1986). 11th ed. English version., Published by The Ministry of Health and Welfare, Japan p. 689.

The United States Pharmacopoeia - The National Formulary. (2003) USP 26. NF 21. Published by the United States Pharmacopoeial Convention, Inc. Rockville, pp

1102 - 1103.

Thomsen, M.H., Bech, D. and Kiel, P. (2004). Manufacturing of stabilized brown juice for L-lysine production – from university lab scale over pilot scale to industrial production. *Chem. Biochem. Eng. Q.* **18**(1), 37 – 46.

Tosaka, O., Hirakawa, H., Takinami, K. amd Hirose, Y. (1978 b) Regulation of lysine biosynthesis by leucine in *Brevibacterium lactofermentum. Agr. Biol. Chem.* **42**(8), 1501.

Tosaka, O., Hirakawa, H., Takinami, K. (1979 b). Effect of biotin levels on L-lysine formation in *Brevibacterium lactofermentum. Agric. Biol. Chem* **43** (3), 491.

Tosaka, O., Ishihara, M., Morinaga, Y. and Takinami, K. (1979 a) Mode of conversion of asperto ß-semialdehyde to L-threonine and L-lysine in *Brevibacterium lactofermentum. Agr. Biol. Chem.* **43**(2), 265.

Tosaka, O. and Takinami, K. (1978). Pathway and regulation of lysine biosynthesis in *Brevibactrium lactofermentum. Agr. Biol. Chem.* **42**(1), 95

Tosaka, O. and Takinami, K. (1986). Lysine. In Biotechnology of amino acid production. Progress in Industrial Microbiology Ed. L. Aida. Vol. 24. Kodansha Ltd. Tokyo, P. 152

Tosaka, O., Takinami, K. and Hirose, Y. (1978 a). L-lysine production by S-(2 aminoethyl)-L-cysteine and a-amino-ß-hydroxyvarelic acid resistant mutants of *Brevibacterium lactofermentum. Agr. Biol. Chem.* **42**(4), 745

Tosaka, O., Takinami, K. and Hirose, Y. (1978 c). Production of L-lysine by leucine auxotrophs derived form AEC resistant mutant of *Brevibacterium lactofermentum. Agric. Biol. Chem.,* **42**(6), 1181.

Tryfona, T and Bustard, M.Y. (2005). Fermentative production of lysine by Corynebacterium glutamicum: transmembrane transport and metabolic flux analysis. *Process Biochemistry,* **40**(2), 499 – 508.

Tsujimoto, N., Gunji, Y., Ogawa-Miyata, Y., Shimaoka, M. and Yasueda, H. (2006). L lysine biosynthetic pathway of *Methylophilus methylotrophus* and construction of an L-lysine producer. *J. Biotechnology.* **124**(2), 327 – 337.

Tucci, A.F. and Ceci, L.N. (1972 a). Homocitrate synthase from yeast. *Arch. Biochem. Biophys.* **153**(2), 742.

Tucci, A.F. and Ceci, L.N. (1972 b). Control of lysine biosynthesis is yeast. *Arch. Biochem. Biophys.* **153**(3), 751.

Uchida, K. and Aida, K. (1979). Taxonomic significance of cell wall acyl type in
    Corynebacterium, Mycobacterium and *Nocardia* group by a glycolate test. *J.*
    Gen. Appl. Microbiol. **25**, 169.

Udeh, K.O. and Achremowich, B. (1993). Optimization of cultivation medium composition
    of an L-lysine producing *Corynebacterium* mutant with the use of response surface
    methodology. *Acta. Microbiol. Pol.* **42**(4), 171.

Velizarov, S.G., Rainina, E.I., Sinitsyn, A.P., Varfolomeev, S.D., Lozinskii, V.I., and
    Zubov, A.L. (1992). Production of L-lysine by free and PVA-cryogel
    immobilized *Corynebacterium glutamicum* cells. *Biotechnol. Lett.* **14**(4), 291.

Velasco, A.M., Leguina, J.I. and Lazcano, A. (2002). Molecular Evolution of the Lysine
    Biosynthetic Pathway. *J. Mol. Evol.* **55**: 445 - 459

Vickery, H.B. and Leavenworth, C.S. (1928). Crystallization of free lysine. *J. Biol.*
    Chem. **76**, 437.

Vogel, H.J.(1963). Lysine pathways as biochemical fossils *Proc. Inter. Congr. Biochem.*
    (5th), Moscow, **3**, 341.

Vrljic, M., Garg, J., Bellmann, A., Wachi, S., Freudl, R., Malecai, M.J., Sahm, H., Kozina,
    V.J., Eggeling, L. and Saier, M.H.Jr. (1999). The LysE superfamily: Topology of the
    lysine exporter LysE of Corynebacterium glutamicum, a paradyme for a noval
    superfamily of transmembrane solute translocators. *J. Mol. Microbiol. Biotechnol..* **1**,
    327 - 336

Vrljic, M., Sahm, H. and Eggeling, L (1996). A new type of transporter with a new type of
    cellular function: L-lysine export from Corynebacterium glutamicum. *Mol.*
    Micriobiol. **22**, 815 – 826.

Wang, J.S., Cheng, W.L., Chang, C.C. and Liu, Y.T. (1994). Improving the L lysine
    productivity of *Brevibacterium sp.* PI-13 *Taiwan Tangye Yanjiuso Yonjiu Huibao,*
    143, 51. Chem. Abs. **121**:228854 j (1994).

Wang, J.S., Kuo, Y C., Chang, C.C. and Liu, Y.T. (1991). Optimization of culture
    conditions for L-lysine fermentation by *Brevibacterium sp.* PI 13. *Taiwan Tangye*
    Yanjiuso Yanjiu Huibao, **134**, 37 Chem. Abs. **118**:253302y (1993).

Wang, J.S., Kuo, Y.C., Chang, C.C. and Liu, Y.T. (1993a). Production of L-lysine by
    Brevibacterium sp. PI-13 in feed-batch fermentation using molasses medium
    Taiwan Tangye Yanjiuso Yanjiu Huibao, **139**, 33 Chem. Abs. **121**:106617p (1994).

Wang, J.S., Kuo, Y.C., Cheng, W.L. and Liu, Y.T. (1993 b). Fermentative production of lysine with raw sugar medium by *Brevibacterium. Sp.* Pl-13., *Taiwan Tangye Yanjiuso yanjiu Huibao.* **139**, 43. Chem. Abs. **121**:106618 q (1994).

Wehrmann, A., Phillipp, B., Sahm, H. and Eggeling, L. (1998). Different modes of diaminopimelate synthesis and their role in cell wall integrity: a study with Corynebacterium glutamicum. *J. Bacteriol.* **180**, 3159 – 3165.

West, E.S., Todd, W.R., Mason, H.S. and Bruggen, J.T.V. (1966). Taxt Book of Biochemistry. 4th ed, The Macmillan Company, London, p. 1356.

White, P.J. (1983). The essential role of diaminopimelate dehydrogenase in the biosynthesis of lysine by *Bacillus spaericus. J. Gen. Microbiol.* **129,** 739 – 749.

Wibowo, D., Budiyanto, G., Jan, L. and Liu J.C. (1992). Kinetic study of lysine fermentaiton in cane molasses base medium. *Proc. Asia Pac. Bio-chem. Eng. Conf.* pp.201.

Windsor, E. (1951). a-aminoadipic acid as a precursor to lysine in *Neurospora. J. Biol. Chem.* **192**, 607.

Wittmann, C. and Becker, J. (2007). The L-lysine story: From Metabolic Pathways to Industrial Production. Microbiology Monographs (Amino Acid Biosynthesis Pathways, Regulation and Metabolic Engineering – Ed: V.E. Wendisch), published by Springer-Verlag, Berlin, Vol – 5, pp 39 – 70.

Won, O.J., Kim. S.J., Cho, Y.J., Park, N.H. and Heung, L.J. (1990). *Corynebacterium glutamicum* mutant for manufacture of lysine. *Fr. Demande FR.* 2,645,172. Chem. Abs. **115**: 7001 c(1991).

Work, E. (1950). A. new naturally occuring amino acid. *Nature.* **165**, 74.

Work, E. (1951). Isolation of a,e-diaminopimelic acid from *Corynebacterium diphtheriae* and *Mycobacterium tuberculosis. Biochem. J.* **49**, 17.

Work, E. (1957). Reaction of ninhydrin in acid solution with straight chain amino acids containing two amino groups and its application to the estimation of a,e-diaminopimelic acid. *Biochem. J.* **67**, 416.

Xu, H., Andi, B.A., Qian, J., West, A.H. and Cook, P.F. (2006). The - aminoadipate pathway for lysine biosynthesis in fungi. Cell Biochem. Biophys. **46**, 43 - 64

Yamamoto, A. (1978). Amino acids- Survey. In. Kirk-Othmer Encycolopedia of Chemical Technology. 3rd. ed., vol. 2. A Wiley-Interscince Publicaion. New York, p. 376.

Yamada, K. and Komagata, K. (1970 a). Taxonomic studies on corynerform bacteria. II. Principal amino acids in the cell wall and their taxonomic significance. *J. Gen.*

Appl. Microbiol. **16**, 103.

Yamada, K. and Komagata, K. (1970 b). Taxonomic studies on coryneform bacteria. III. DNA base composition of coryneform bacteria. *J. Gen. Appl. Microbiol.* **16**, 215.

Yamada, K. and Komagata, K. (1972 a). Taxonomic studies on coryneform bacteria. IV. Morphological, Cultural, Biochemical, and Physiological characteristics. *J. Gen. Appl. Microbiol,* **18**, 399

Yamada, K. and Komagata, K. (1972 b). Taxonomic studies on coryneform bacteria. V. Classification of coryneform bacteria *J. Gen. Appl. Microbiol.* **18**, 417.

Yinghua, F., Suowei, X., Xingzhen, W., Lingming, S., Zeyu, X. and Guoliang, S. (1992). Studies on L-lysine producing strain *Brevibacterium flavum* Au-112. *Gongye Weishengwu,* **22**(1), 3., Chem. Abs. **116**:233823 z (1992).

Yenekura, H., Hirao, T., Azuma, T. and Nakanishi, T. (1988). Lysine manufacture with *Corynebacterium* mutants. *Fr. Demande Fr.* 2,601,035. Chem. Abs. **109**:36624q (1988).

Yoshihara, Y., Kawahara, Y. and Ishii, T. (1990). Fermentative manufacture of L-lysine with *Brevibacterium* and *Corynebacterium* species. *Japanese Patent.* 02234,686. Chem. Abs. **114**:120274m (1991).

Zabriskie, T.M. and Jackson, M.D. (2000). Lysine biosynthesis and metabolism in fungi. *Nat. Prod. Rep.,* **17**, 85 – 97.

Zaitseva, Z.M., Kapitonova, O.N. and Oranskaya, M.S. (1973). Physiological and cytological properties of two strains of *Micrococcus glutamicus* during lysine biosynthesis. *Priki. Biokhim Mikrobiol.* **9**(1), 60 Chem. Abs. **78**:133187 y (1973).